Quantum Cosmos Theory

Fractal Cosmos III

Stephen Blaha Ph. D.
Blaha Research

MMXXV

ISBN: 979-8-9986071-6-5

Rev. 00/00/01 October 4, 2025

To Margaret

Some Other Books by Stephen Blaha

SuperCivilizations: Civilizations as Superorganisms (McMann-Fisher Publishing, Auburn, NH, 2010)

Unification of God Theory and Unified SuperStandard Model THIRD EDITION (Pingree Hill Publishing, Auburn, NH, 2018).

The Exact QED Calculation of the Fine Structure Constant Implies ALL 4D Universes have the Same Physics/Life Prospects (Pingree Hill Publishing, Auburn, NH, 2019).

Passing Through Nature to Eternity ProtoCosmos, HyperCosmos, Unified SuperStandard Theory (Pingree Hill Publishing, Auburn, NH, 2022).

HyperCosmos Fractionation and Fundamental Reference Frame Based Unification: Particle Inner Space Basis of Parton and Dual Resonance Models (Pingree Hill Publishing, Auburn, NH, 2022).

The Cosmic Panorama: ProtoCosmos, HyperCosmos, Unified SuperStandard Theory (UST) Derivation (Pingree Hill Publishing, Auburn, NH, 2022).

God and and Cosmos Theory (Pingree Hill Publishing, Auburn, NH, 2023).

Newton's Apple is Now The Fermion (Pingree Hill Publishing, Auburn, NH, 2023).

Cosmos Theory: The Sub-Particle Gambol Model (Pingree Hill Publishing, Auburn, NH, 2023).

Cosmos-Universe-Particle-Gambol Theory (Pingree Hill Publishing, Auburn, NH, 2024).

Structure and Dynamics of Cosmos Theory and the Unified SuperStandard Theory (Pingree Hill Publishing, Auburn, NH, 2024).

Black Holes, White Holes, and Superluminal Starship (Pingree Hill Publishing, Auburn, NH, 2025).

The Sequence Theory Foundation of Cosmos Theory and Gambol Theory (Pingree Hill Publishing, Auburn, NH, 2025).

BEYOND TIME AND SPACE Fermion Strata & Consistency Conditions for Sequences, Cosmos Spaces and Gambols (Pingree Hill Publishing, Auburn, NH, 2025).

The C o s m o s as a Physics Organism (Blaha Research, Auburn, NH, 2025)

The Root of Reality (Blaha Research, Auburn, NH, 2025)

Fractal Cosmos (Blaha Research, Auburn, NH, 2025)

Fractal Cosmos II (Blaha Research, Auburn, NH, 2025)

Available on bn.com, Amazon.com, Amazon.co.uk and other international web sites as well as at better bookstores.

CONTENTS

FIGURES and TABLES

Introduction

The Cosmos Fractal Curve is the foundation of the Cosmos. This book extends the fractal concept to a *quantum* fractal curve. It makes the Cosmos Quantum as well as fractal. The book describes the construction of a quantum fractal curve and applies it to create the Quantum Cosmos Fractal Curve. This probabilistic curve is then extended to include the dimension arrays of Cosmos spaces and their associated space-times. The result is a flattened double helix similar to the DNA of life.

One result of this development is a view of the Dark sector of the universe. Our universe may be a quantum superposition of a universe with no Dark sector and another universe with a Dark sector. We do not see the Dark sector except through gravitation. All interactions between the superposed universes are suppressed. The result is what we see presently. Only gravitational effects.

The book presents a model of a quantum fractal curve and shows how to calculate its probabilities. Interestingly, the probability of the Cosmos spaces that were previously found is a comfortable 72.36% – good odds.

The genesis of the Master Equation is considered. And the quantum creation of universes from a quantum Cosmos space state is developed to the point where a tangible universes result through an inverse Z-order Lebesgue transformation.

The further creation of a particle within the created universe is also described in detail.

The question of the interior of elementary particles is considered. A Scaling Planckian distribution is defined that is applied to deep inelastic electron-proton scattering. An excellent match is found. This match leads to the view that a fundamental particle is composed of 844242 Gambol Fermions that undergo internal interactions that yield a fractal dynamics. An analogy to the shoreline of the ocean is made where the fractal nature of the shoreline is generated by the combined dynamics of the water waves.

The author shows that the entire Cosmos from spaces to particles is fractal.

1. The Essence of the Cosmos

Cosmos Theory has many interrelated features. In this book we develop its components in the form of a series of construction steps realizing the construction is descriptive in nature. The actual construction is instantaneous. Since there is no time at the point of construction we may view it as an approximate description (simulation) of Reality that has the goal of providing an understanding of the Cosmos. To the extent the description is faithful it also has a predictive power – an investigation of the implications of its components can be expected to yield new insights into the features of Reality.

We will consider the components of the Cosmos in a sequential order while realizing that they exist spontaneously and simultaneously.

1. The "initial point" of the Cosmos is a mathematical, zero dimension point. We see it as the limit of a fluctuation of zero probability occurring in an infinite time period – if there were a time parameter. One might view this situation as non-trivial since the infinite time limit of a "zero" fluctuation probability is symbolically "$\infty 0$".
2. A series of construction steps follow that generate a two dimension space-filling fractal curve called the Cosmos Fractal Curve. There is no space, time, mass or energy at this "stage."
3. The steps in the construction of the Cosmos Curve are numbered. The nth step has a piece-wise linear segment of length 2^n. For even n we associate a HyperCosmos space with a dimension array of size $2^n = 2^{r+4}$. For odd n we associate a 2nd Kind HyperCosmos space with a dimension array of size $2^n = 2^{r+3}$. The consequence of these assignments is a representation of the Cosmos Curve as double stranded with one strand (for even n values) containing the HyperCosmos spaces and the other strand (for odd n values) containing 2nd Kind HyperCosmos spaces.
4. Each construction step has an associated space and dimension array, and a space-time $D(2^r)$ with $n = r + 4$ (HyperCosmos space) or $n = r + 3$ (2nd Kind HyperCosmos space).
5. These strands, when represented mathematically, reveal a quantum interpretation of Cosmos spaces that we pursue below in our consideration of Quantum Theory in the two dimension space.
6. The Cosmos Curve becomes a two dimension unit square that expands by scaling to become a two dimension space.
7. A time coordinate can be defined in this space as the distance along the fractal curve. We map the curve with its irregularities to a time coordinate represented by a continuous "straight line." The time coordinate has an infinite range.

8. Having a well-defined time we can introduce a Quantum Theory in the two dimension space with a time coordinate that results in a Schrödinger=like equation for the scaling Cosmos Master Equation.
9. The "energy" eigenvalues of the Master Equation are the dimension array sizes of the HyperCosmos and 2nd Kind HyperCosmos spaces.
10. From the Cosmos spaces one obtains the Unification spaces and the Fundamental Reference Frames (FRF).
11. The Quantum Theory of the Cosmos is then explored.
12. Up to this point we have obtained the structure of Cosmos Theory but not universes and particles.
13. Now a Parent universe appears as part of a consistency condition that limits the number of HyperCosmos spaces. The number of Physical HyperCosmos spaces is determined and equals the number of Physical 2nd Kind HyperCosmos spaces. The space-time dimension of the Parent universe is fixed by the consistency condition. Its dimension array size is also determined.
14. Subuniverses may be created dynamically within the Parent universe.
15. The dimension arrays give the fundamental fermion spectrum and the symmetry group patterns.
16. The Master Equation has associated sequences for spaces and universes that can be directly applied to spaces, fermion masses and coupling constants values.
17. The various components of Cosmos Theory are then mapped to a fractal representation conceptually based on the Cosmos Fractal Curve.
18. Models are created for the internal structure of fundamental fermions and hadrons. These models use a Scaling Planckian Distribution described in chapters 6 and 7.

The above steps are sequential. However they must be viewed as instantaneous and occurring outside of any time. An ever present Cosmos! Within the Parent universe(s) a time exists and the question of timing becomes relevant for subuniverses.

In the following chapters we consider details of items in the above list. There are additional features and details in the previous Fractal Cosmos volumes and in previous books by the author over the past seven years. The Unified SuperStandard Theory (UST) is described in books written since the year 2000.

2. Double Helix Fractal Cosmos Curve

This chapter[1] develops the Fractal Cosmos Curve that leads to a (flattened) double helix Cosmos Curve. It presents the Fractal Cosmos Curve based on the fractal notation and arithmetic described in Appendix 2-A.

2.1 The Fractal Curve of the Cosmos – Double Threaded

The Cosmos Curve begins with a zero dimension point, and grows to a one dimension piecewise linear line segment of unit length by a simple aggregation of infinitesimal elements based on the identity

$$\sum_{k=\infty}^{1} 2^{-k} = 1 \tag{2.1}$$

It starts from a point where $k = \infty$, initially adding infinitesimal increments, which then sum to a unit line segment. The order of the summation begins with the infinitesimal and proceeds to the larger summands.

Then the unit piecewise linear line segment grows through a construction consisting of steps to form a fractal line. At the nth stage of the construction the growing line has the length 2^n. This line is compressed to fit within a unit square of sides 1 by 1. The line becomes fractal in the limit $n \rightarrow \infty$. The resulting fractal line is known as the Hilbert curve and, in the present context, the Cosmos Fractal Curve.

2.1.1 The Steps of the Construction

The steps of the construction process display a feature that relates to the distinction between HyperCosmos spaces and 2nd Kind HyperCosmos spaces.[2] The feature follows from our requirement that the dimension r be an even integer.[3] There is an important *Physical* distinction between these types of spaces. HyperCosmos spaces' dimension arrays have both a Normal sector and a Dark sector while 2nd Kind HyperCosmos spaces' dimension arrays have only a Normal sector. The difference is a factor of 2 in the sizes of dimension arrays. Our UST is an example since it has an $r = 4$ HyperCosmos space dimension array with $2^{4+4} = 256$ components and has both a Normal and Dark sector. A *different* choice of a 2nd Kind HyperCosmos dimension array with $2^{4+3} = 128$ components would only be a Normal sector as we discussed in earlier books.

---

[1] Largely based on Blaha (2025f).

[2] A HyperCosmos space has both a Normal and a Dark sector. A 2nd Kind HyperCosmos space only has a Normal sector. As a result their dimension array sizes differ by a factor of 2.

[3] The value of r is even to prevent double counting of spaces since an r+1 dimension array is the same as an r dimension array.

For *even* n we have the piecewise linear segment of length 2^n that corresponds to a HyperCosmos space of dimension r whose dimension array size is 2^{r+4}. We can map the linear segment length to the dimension array size by linearizing the dimension array square into a sequence of dimensions. The result is

$$2^n = 2^{r+4} \tag{2.2}$$

Then

$$n = r + 4 \tag{2.3}$$

In terms of fractal dimensions[4] we find

$$D(2^n) = D(2^{r+4}) \tag{2.4}$$

for a HyperCosmos dimension array map.

For *odd* n we have a piecewise linear segment of length 2^n. In order to preserve the oddness of n we must relate it to a 2nd Kind HyperCosmos space of dimension r whose dimension array size is 2^{r+3}. Again we can map the linear segment length to the dimension array size by linearizing the dimension array into a sequence of dimensions. The result is

$$2^n = 2^{r+3} \tag{2.5}$$

with

$$n = r + 3 \tag{2.6}$$

or

$$D(2^n) = D(2^{r+3}) \tag{2.7}$$

for a 2nd Kind HyperCosmos space. The odd n is matched by the odd value of r+ 3 of 2nd Kind HyperCosmos dimension arrays since r is an even integer.

2.1.2 The Sequence of Construction Steps

Given the above comments, we can define the construction steps in terms of HyperCosmos spaces as:

$$D(2^n) = \delta_{n,even} D(2^{r+4}) + \delta_{n,odd} D(2^{r+3}) \tag{2.8}$$

or

$$D(2^n) = \cos^2(n\pi/2)\, D(2^{r+4}) + \sin^2(n\pi/2)\, D(2^{r+3}) \tag{2.9}$$

where even values of n give HyperCosmos $D(2^n) = D(2^{r+4})$ and odd values give 2nd HyperCosmos $D(2^n) = D(2^{r+3})$. Eq. 2.8 shows that the construction steps of the fractal curve alternate between HyperCosmos space and 2nd Kind HyperCosmos space equivalents. We may view the construction process as alternating construction steps of HyperCosmos space, and 2nd Kind HyperCosmos space, piecewise linear segments.

Fig. 2.6 displays a part of the Cosmos Fractal Curve showing the "oscillation" in n between Cosmos and 2nd Kind Cosmos steps.

We now let $n\pi/2 \rightarrow$ the angle variable θ and consider

---

[4] The D() notation is described in the 2-A Appendix on Fractal notation and arithmetic.

$$D(n, n', r, \theta) = \cos^2(\theta)D(\text{HyperCosmos})|_{n=n'=r+4} + \sin^2(\theta)D(2^{nd}\text{ Kind HyperCosmos})|_{n\pm1=n'=r+3\pm2} \quad (2.10)$$

We examine the change in D(n, n′, r, θ) as the angle θ changes from 0 through π/2. The angle θ *measured from HyperCosmos strand points* of the helix range from the HyperCosmos strand $D(\text{HyperCosmos})|_{n=n'=r+4}$ to two possible 2nd Kind HyperCosmos strand points where $D(2^{nd}\text{ Kind HyperCosmos})|_{n\pm1=n'=r+3\pm2}$. One point has the value

$$D(2^{nd}\text{ Kind HyperCosmos})|_{n+1=n'=r+3+2} = D(2^{nd}\text{ Kind HyperCosmos})|_{n+1=n'=r+5=r'+3} \quad (2.11)$$

where n′ = n + 1 and the dimension changes to r′ = r + 2 signaling an "up move" on that strand. The other point has the value

$$D(2^{nd}\text{ Kind HyperCosmos})|_{n-1=n'=r+3-2} = D(2^{nd}\text{ Kind HyperCosmos})|_{n-1=n'=r+1=r'+3} \quad (2.12)$$

where n′ = n – 1 and the dimension changes to r′ = r – 2 signaling a "down move" on that strand.

A similar relation can be created for an angle θ going from 2nd Kind HyperCosmos points to HyperCosmos points. We examine the change in $D_2(n, n', r, \theta)$ as the angle θ changes from 0 through π/2.

$$D_2(n, n', r, \theta) = \cos^2(\theta)D(2^{nd}\text{ Kind HyperCosmos})|_{n'=n=r+3} + \sin^2(\theta)D(\text{HyperCosmos})|_{n'=n\pm1=r+4\pm2} \quad (2.13)$$

Here the angle θ *measured from the 2nd Kind HyperCosmos strand points* of the helix range from the 2nd Kind HyperCosmos strand $D(2^{nd}\text{ Kind HyperCosmos})|_{n'=n=r+3}$ to two possible HyperCosmos strand points where $D(\text{HyperCosmos})|_{n'=n\pm1=r+4\pm2}$. One point has the value

$$D(\text{HyperCosmos})|_{n'=n+1=r+4+2} = D(\text{HyperCosmos})|_{n'=n+1=r+6=r'+4} \quad (2.14)$$

where n′ = n + 1 and the dimension changes to r′ = r + 2 signaling an "up move" on that strand, and the other point has the value

$$D(\text{HyperCosmos})|_{n'=n-1=r+4-2} = D(\text{HyperCosmos})|_{n'=n-1=r+2=r'+4} \quad (2.15)$$

where n′ = n – 1 and the dimension changes to r′ = r – 2 signaling a "down move" on the strand.

Eqs. 2.10 and 2.13 each generate a double helix in two dimensions – a flattened double helix. See Fig. 2.7 for the eq. 2.10 based helix.

2.1.3 Quantum Interpretation of Oscillation in θ

Eq. 2.10 appears rather like a quantum probability expression with

$$\text{Probabilities: } \cos^2(\theta) \text{ and } \sin^2(\theta) \quad (2.16)$$

$$\text{Amplitude factors: } e^{i\varphi}\cos(\theta) \text{ and } e^{-i\varphi}\sin(\theta) \quad (2.17)$$

One might interpret eq. 2.10 as the relative probability of having a HyperCosmos space or a 2nd Kind HyperCosmos space for a given value of space-time dimension r:

$$AD(n, n', r, \theta) = \cos(\theta)AD(\text{HyperCosmos})|_{n=n'=r+4} + \sin(\theta)AD(2^{nd}\text{ Kind HyperCosmos})|_{n\pm1=n'=r+3\pm2} \tag{2.18}$$

where AD(…) represents the amplitude. For example

$$D(\text{HyperCosmos})|_{n=n'=r+4} = |AD(\text{HyperCosmos})|_{n=n'=r+4}|^2 \tag{2.19}$$

An alternative expression of the quantum amplitudes uses wave functions:

$$\Psi = e^{i\varphi}\cos(\theta)\psi_H + e^{-i\varphi}\sin(\theta)\psi_{H2} \tag{2.18'}$$

where ψ_H and ψ_{H2} are wave functions for a HyperCosmos space and a 2nd Kind HyperCosmos space respectively. *Eq. 2.18′ creates a quantum field Ψ for each step in the construction of the Cosmos Fractal Curve. This field gives an amplitude for the creation of the Cosmos dimension array at that step.*

We therefore raise the possibility of a Quantum interpretation of the Fractal Cosmos Curve in chapter 3. A potential problem is the apparent lack of a Quantum interpretation of the space of the fractal curve, if any, within which the curve may reside. This issue is partly mitigated by the two dimension space generated by the space-filling Cosmos Fractal Curve with the introduction of a time parameter as the distance along the fractal curve, and by the Schrödinger-like form of the Master Equation that can be defined. The distance along the fractal curve can be viewed as a time order embodied in Fractal Curve construction.

Since the curve and its implicit two dimension space, and the two dimension space of the Master Equation, and the Master Equation itself, and the Cosmos spaces spectrum are all *structural* with no basis in Reality (since they have no mass or energy), they can be considered to be infinitely existing components. True times only exist within universes.

Eq. 2.18 can be interpreted as a probability amplitude that exists at the point of creation of a universe of space-time dimension r. It implies that a universe may be created having a HyperCosmos space structure with a dimension array of size 2^{r+4} or a 2nd Kind HyperCosmos space with a dimension array of size 2^{r+3} or a superposition of these space types. When a universe is instantiated it has one of these three possibilities. The difference between a HyperCosmos space structure and a 2nd Kind HyperCosmos space structure is the presence of a Dark sector for the first, and the absence of a Dark sector for the second respectively, due to the factor of 2 difference in dimension array sizes.

2.1.4 Quantum Interpretation of a Dark Sector in Our Universe

It is not certain whether a Dark sector exists for our universe. Yet there is indirect evidence for a Dark sector based primarily on gravitation studies. There is no convincing direct evidence. This dichotomy raises the possibility that our universe is a superposition of these space types. Then the Dark sector does not "exist" but the quantum effects of the superposition generate indirect "evidence" for a Dark sector based on quantum based gravitation effects.

2.2 Double Helix Fractal Curve

We note that the curves of values generated by the variations in θ have the appearance of a helix as shown in Fig. 2.7 for several reasons.

Firstly, the variation from $\theta = 0$ continuously to $\theta = \pi/2n$ takes $D(2^n, \theta)$ from $D(2^{r+4})$ to $D(2^{r+3})$. This change corresponds to a change from n to n + 1 as shown in Fig. 2.7 with solid curving "helix" lines (the "front" view of the helix).

Secondly, the variation from $\theta = \pi/2n$ continuously to $\theta = 0$ takes $D(2^n, \theta)$ from $D(2^{r+3})$ to $D(2^{r+4})$. This change corresponds to a change from n to n + 1 as shown in Fig. 2.7 with broken curving "helix" lines (the "rear" view of the helix).

Thirdly, the absence of spatial dimensions for the Cosmos Curve causes what might have been a 3-D truly helix curve to be a flattened helix between values as shown in Fig. 2.7. The construction step n may be viewed as a dimension of a space associated with the piecewise linear curves.

The sequence of construction of the Cosmos Fractal Curve alternates between HyperCosmos and 2nd Kind HyperCosmos space dimensions. The even and odd n construction steps support a separation of the Cosmos Curve into two strands that are interconnected step by step.

As a result the Cosmos Fractal Curve is the equivalent of a Double Helix curve. It thus is an analogue of DNA – the structure of organic life. We will call it the *Cosmic DNA* curve. See Fig. 2.8.

One may view the correspondence between the (flattened) double helix form of the Cosmos DNA Curve in two dimensions, and the three dimension double helix form of DNA, as reflecting a similar "solidity" of structure. The ubiquity of DNA in organic life reflects its essential strength as a Physical structure. DNA and the fractal Cosmos Curve are both long strands with an inherent stability.[5]

Our results support our contention that Cosmos Theory embodies a type of Physical Organism. The ubiquitous presence of fractals leads us to view the Cosmos as striated.

2.3 Extension of the Dimension Array Concept to a 2nd Space and a Space-time

By viewing a dimension array size as a fractal dimension, the possibility of extending our view of the Cosmos Curve appears. For each construction step n we can view the dimension array size[6] (2^{r+4} or 2^{r+3}) as the fractal (topological) dimension of its *associated Cosmos space*.

---

[5] This author was perhaps the first to investigate the electronic structure of DNA using electron spin resonance (ESR) techniques in 1964 in the laboratory of Professor Young Wook Kim, Wayne State University.(Detroit).

[6] At step n we have n = r + 4 for n even, and n = r + 3 for n odd.

For each n and r a new second space also appears. This space has r space-time dimensions. There is thus an implicit coordinate space associated with each step n and its associated Cosmos space. One can view the Cosmos fractal construction process as also *generating both a Cosmos space and a space-time at each step n.*

The creation of a universe of the type of a Cosmos space of dimension r is associated with a space-time of dimension r with dimensions allocated within the space's dimension array. Cosmos spaces have both dimension arrays *and* space-times that are used in the formation of universes. *A universe of space type r is fractal as well and has the dimension r.*

2.4 Second Kind HyperCosmos Spaces

The appearance of 2nd Kind HyperCosmos space steps in the Cosmos fractal construction corresponds to their appearance in the unification of Cosmos spaces. Fig. 2.4 shows the set of unification spaces of Cosmos Theory. This figure and the following Fig. 2.5 were presented in previous books by the author.

Fig. 2.5 shows the dimension array of the 42 dimension Full HyperUnification space. It has three parts: an A part for HyperCosmos spaces unification, a B part for 2nd Part HyperCosmos spaces unification, and a C part for a remaining dimension array.

Dimension array A part is 5,592,400×5,592,400 dimensions
Dimension array B part is 2,796,200×2,796,200 dimensions
Dimension array C part is 8×8 dimensions

The 2nd Kind HyperCosmos spaces that appear here (and in parts of previous books) correspond directly with the structure of the Cosmos Fractal Curve presented above. Thus we have a consistency between the Cosmos Fractal Curve construction and the analysis of the unification structure of Cosmos Theory spaces.

This chapter has shown the detailed relation of Cosmos Curve construction steps to HyperCosmos and 2nd Kind HyperCosmos spaces. It has also associated a Cosmos space with (topological) dimension, and also with a space-time for each construction step n.

THE HYPERCOSMOS SPACES SPECTRUM

Blaha Space Number $N = o_s$	Cayley-Dickson Number n	Cayley Number Cn d_c	Dimension Array column length d_{cd}	Dimension Array Size $d_{dN}=d_{dr}$	Space-time-Dimension r	CASe Group $su(2^{r/2},2^{r/2})$ CASe
0	10	1024	2048	2048^2	18	su(512,512)
1	9	512	1024	1024^2	16	su(256,256)
2	8	256	512	512^2	14	su(128,128)
3	7	128	256	256^2	12	su(64,64)
4	6	64	128	128^2	10	su(32,32)
5	5	32	64	64^2	8	su(16,16)
6	4	16	32	32^2	6	su(8,8)
7	**3**	**8**	**16**	$\mathbf{16^2}$	**4**	**su(4,4)**
8	2	4	8	8^2	2	su(2,2)
9	1	2	4	4^2	0	su(1,1)

Figure 2.1. The HyperCosmos space spectrum.

THE Limos SECTOR SPACES SPECTRUM

Blaha Space Number $N = O_s$	Cayley-Dickson Number n	Cayley Number d_c	Row Size d_{cr}	Dimension Array Size $d_{dN}=d_{dr}$	Space-time-Dimension r	CASe Group $su(2^{r/2},2^{r/2})$ CASe
10	0	1	2	2^2	-2	U(1)
11	-1	½	1	1^2	-4	U(½)
12	-2	¼	½	$½^2$	-6	U(¼)
13	-3	1/8	¼	$¼^2$	-8	U(1/8)
14	-4	1/16	1/8	$1/8^2$	-10	U(1/16)
				•		
				•		
				•		
∞	−∞	0	0	0	−∞	U(0)

Figure 2.2. The Cosmos Theory Limos spaces spectrum for Gambol Theory. See Blaha (2023e).

HYPERCOSMOS OF THE SECOND KIND SPACES SPECTRUM

Blaha Space Number $N = O_s$	Cayley-Dickson Number n	Cayley Number d_c	Dimension: Array size $d_{dN2} = d_{dr2}$	Space-time-Dimension r	CASe Group $su(2^{r/2}, 2^{r/2})$ CASe
0	10	1024	1024 × 2048	18	su(512,512)
1	9	512	512 × 1024	16	su(256,256)
2	8	256	256 × 512	14	su(128,128)
3	7	128	128 × 256	12	su(64,64)
4	6	64	64 × 128	10	su(32,32)
5	5	32	32 × 64	8	su(16,16)
6	4	16	16 × 32	6	su(8,8)
7	**3**	**8**	**8 × 16**	**4**	**su(4,4)**
8	**2**	4	4 × 8	2	su(2,2)
9	**1**	2	2 × 4	0	su(1,1)

Figure 2.3. Part of the HyperCosmos of the Second Kind space spectrum.

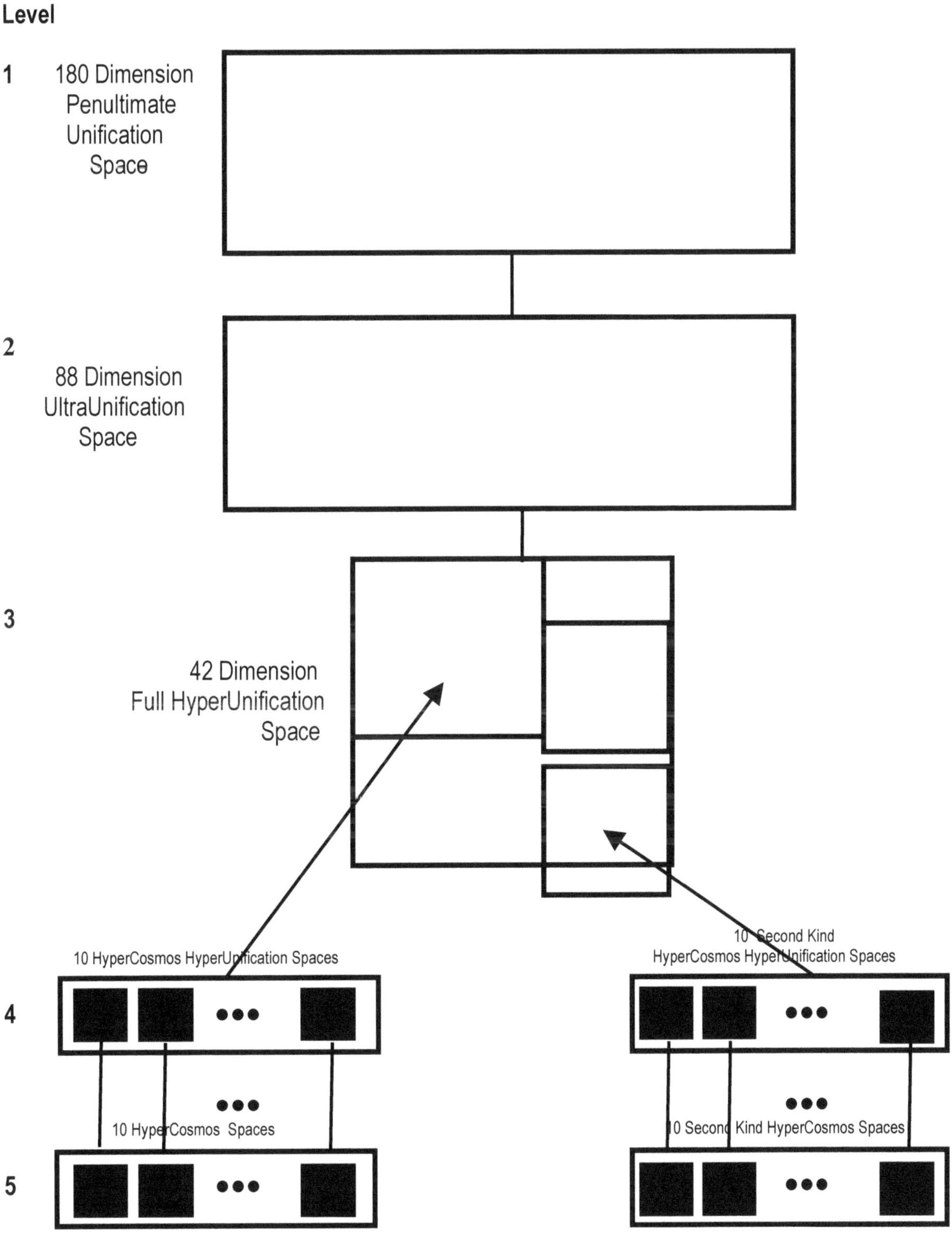

Figure 2.4. Diagram of the five levels of the Cosmos spaces unification. They contain 43 spaces.

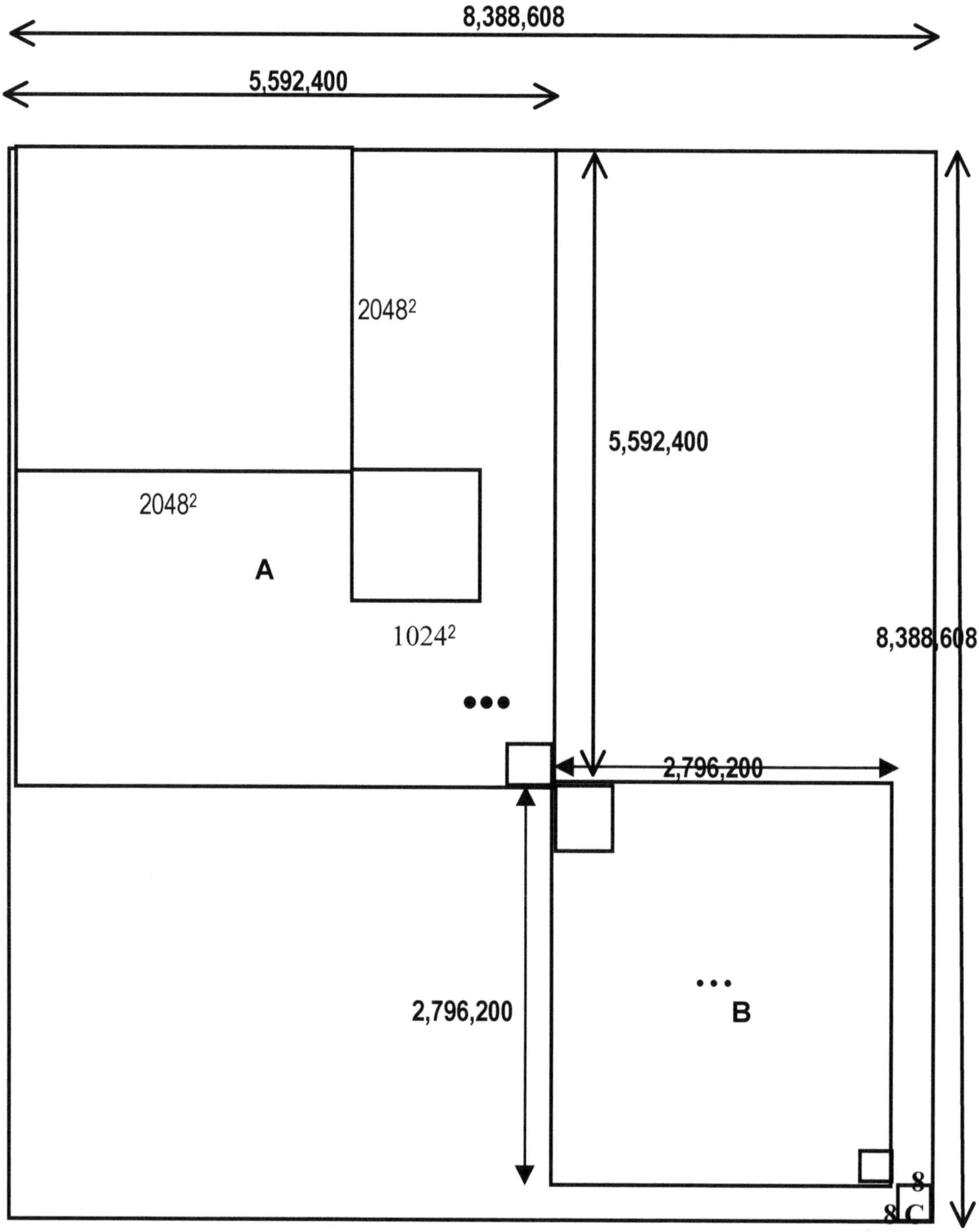

Figure 2.5. Form of the 42 space-time dimnesion of the Full HyperUnification space. All blocks are square. The figure is not drawn to scale.

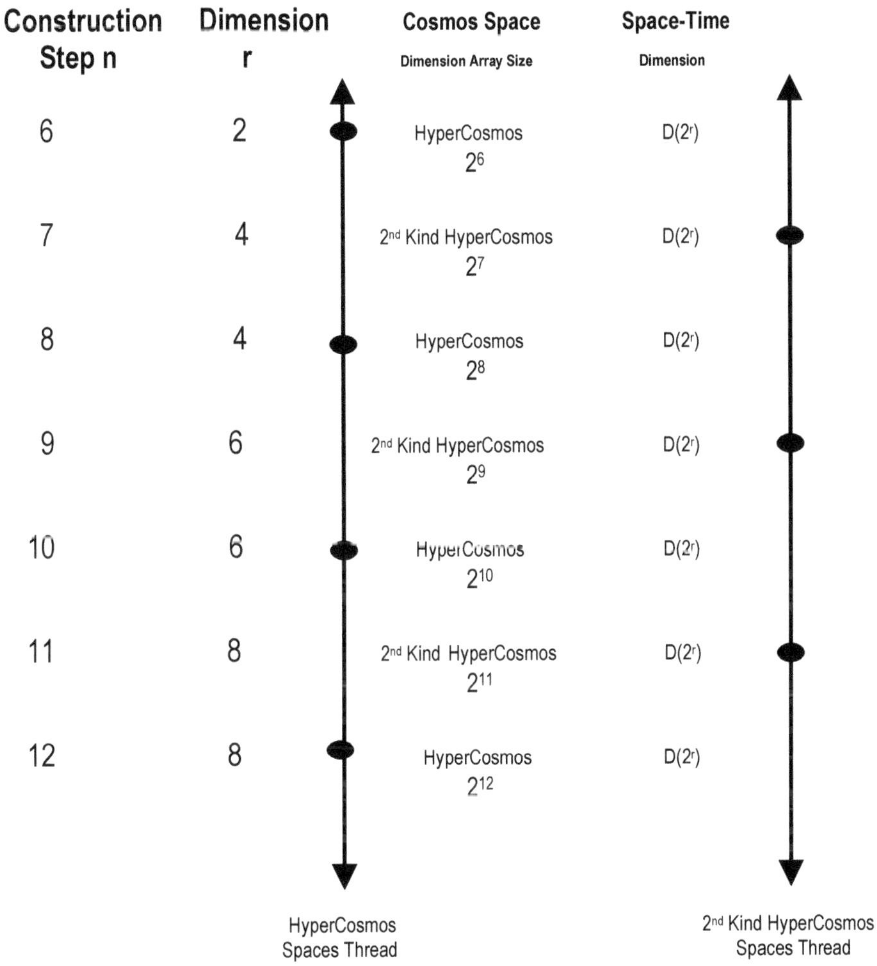

Figure 2.6. Section of the Cosmos Curve showing the two threads of Cosmos spaces and space-times. The space-times $D(2^r)$ for each n is based on eq. 2.1. The piecewise linear line length is 2^n. At each construction step n there is the value of the linear line, its Cosmos space equivalent, and the space-time equivalent. Note the repetitions of space-time dimensions is correct since it reflects the presence of both HyperCosmos and 2nd HyperCosmos spaces in the two threads. The "dots" indicate the type and presence of each Cosmos space on each thread for that value of n. Note the alternation between threads as n increases. The powers of 2 are the dimension array sizes.

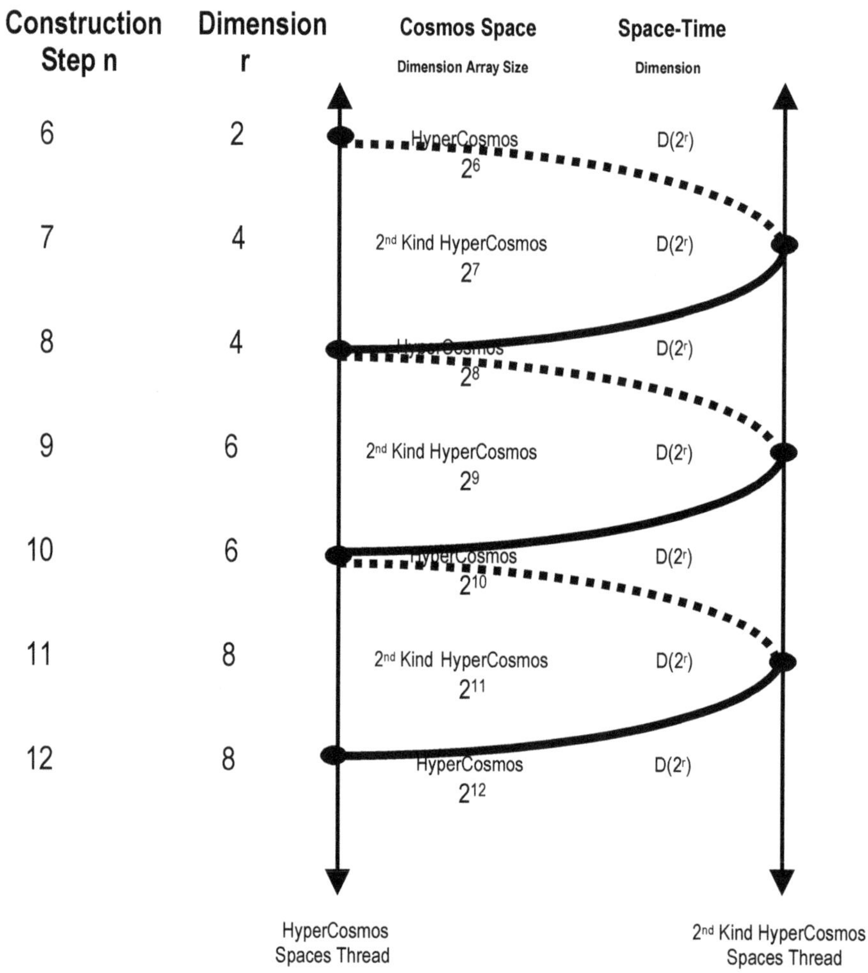

Figure 2.7. A section of the Cosmos Curve showing a flattened double helix form (with the helix mapped to two dimensions.) The two threads of Cosmos spaces and space-times with the "helix" connection between strands are due to the variation in θ. The connecting curved lines form a helix when the 0 spatial dimension of the cutve is considered – an analogue of a double helix similar to DNA. The variation from $\theta = 0$ continuously to $\theta = \pi/(2n)$ takes $D(2^n, \theta)$ from $D(2^{r+4})$ to $D(2^{r+3})$. This change corresponds to a change from n to n + 1 as shown above with solid curving "helix" lines (the "front" view of the helix). The variation from $\theta = \pi/(2n)$ continuously to $\theta = 0$ takes $D(2^n, \theta)$ from $D(2^{r+3})$ to $D(2^{r+4})$. This change corresponds to a change from n to n + 1 as shown above with broken curving "helix" lines (the "rear" view of the helix). The curved lines are not drawn to scale.

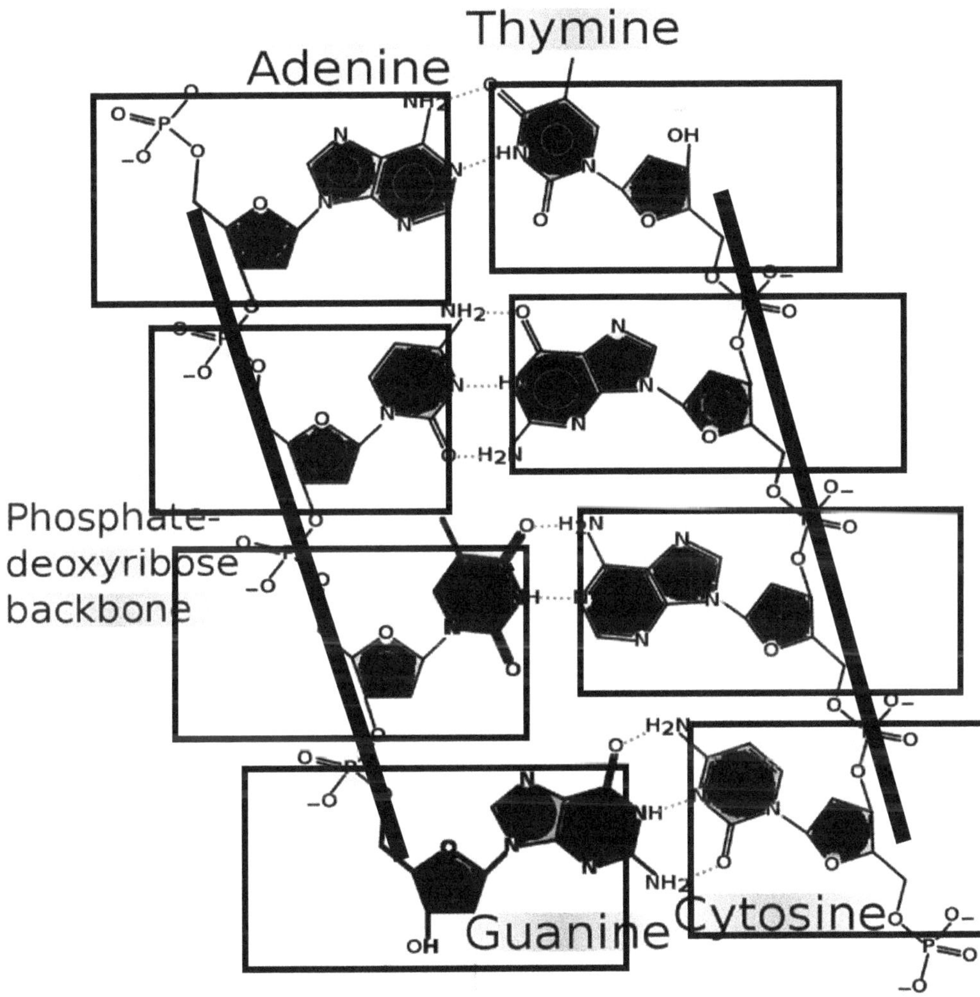

Figure 2.8. The Cosmos double helix structure superimposed on a four base pair fragment of double helix DNA. The correspondence is evident from Fig. 5.3 of Blaha (2021e).

Appendix 2-A. Brief View of Fractals

The dimension D of simple and fractal curves is called the Hausdorff dimension. It can be an integer, as it is for simple spaces such as 3-dimension space where D = 3. It can be a non-integer value for a space filling curve such as the Hilbert curve where the line occupies a two dimension space.

2-A.1 Hausdorff Dimension

The Hausdorff dimension D of a curve is

$$D = \log(N)/\log(\varepsilon) = \log_{\varepsilon}(N) \qquad (2\text{-A}.0)$$

where the numerator = log N equals the logarithm of the number of copies of the shape formed from each copy of the previous iteration of the curve under construction. The denominator = log ε is the logarithm of the factor by which copies are scaled downward from the previous iteration.

For example, a Sierpinski triangle (double sided) creates 3 copies resulting in N = 3 copies from the previous iteration. The scale is ε = 2 based on the doubling of the side. As a result:

$$D = \log(3)/\log(2) = \log_2(3) = 1.585$$

or

$$2^D = 3$$

For the Cosmos Curve and the Hilbert curve we find N = 4 and ε = 2 with the resulting fractal dimension 2:

$$D = \log(4)/\log(2) = 2$$

2-A.2 Fractal Cosmos Spaces and Fractal Dimension Arithmetic

The Cosmos Fractal Curve is described (created) in a series of steps – *conceptually*. In Reality it exists instantaneously. There is no point in time for its creation since time does not exist "before" the Cosmos Fractal Curve. So the "creation" process that we describe has the sole purpose of providing an understanding of the Cosmos. The Cosmos Fractal Curve fills a two dimension unit square. It has fractal dimension 2.

At each step of the Cosmos Fractal Curve construction there is a *numerical* value 2^n – the length of the piecewise linear curve used in construction. The number denoted n has only been used as the linear curve length until now. We will now expand its definition based on its value being related to the number of components (dimensions) in a Cosmos space.

A HyperCosmos space of dimension r has a dimension array with 2^{r+4} components. We match the step to a HyperCosmos space by setting n = r + 4. In the case of a 2nd Kind HyperCosmos space we set n = r + 3 since the space has a dimension array with 2^{r+3} components.

We have extended the piece-wise linear curve number step n to include an associated Cosmos space as above. The definition of fractal dimension (including topological dimensions) may be applied to Cosmos spaces by setting N in eq. 2-A.0 with

$$N = 2^n$$

and ε = 2. The result is

$$D(2^n) = \log(2^n)/\log(2) = \log_2(2^n) = n \qquad (2\text{-A.0a})$$

which gives the Fractal (topological) dimension of the associated space.

Each Cosmos space now has an associated fractal of fractal dimension $D(2^{r+4})$ = r + 4 (HyperCosmos spaces) or $D(r^{r+3})$ = r + 3 (2nd Kind HyperCosmos spaces). We are justified in associating construction step n with the Cosmos space of dimension n since there is a 1:1 map between construction steps and Cosmos spaces.

In addition to associating a HyperCosmos space with each construction step we will also associate a space-time[7] of dimension r = n - 4 or r = n - 3 with construction step n.[8] These space-times do not exist until universes are created. They become the space-times of the universes. As an associate of a construction step, a space-time is an artifact of Cosmos structure and thus not Physically instantiated in the structure. See Fig. 1.6 for a diagram of a section of the Cosmos Fractal Curve together with its associated spaces.

At each construction step there is the length of the piecewise linear curve, a Cosmos space and a space-time.

We will find it necessary to "add", "subtract", "multiply", and "divide" fractals as described later in this appendix. When we perform arithmetic operations they will be applied to fractal dimensions although they will be exemplified with Cosmos dimension arrays. We define fractal dimensions from Cosmos dimension array sizes. Since the fractal dimensions are based on logarithms as in eq. 2-A.1 below the arithmetic operations on fractal dimensions will be based on logarithms of dimension array sizes. Thus, for example, fractal dimension addition is based on products of dimension array sizes. See eq. 2-A.3 below.

Fractal dimension arithmetic (in these discussions) will be based on dimension array sizes and the nature of logarithms, which lead to a different form for arithmetic operations.

2-A.3 Fractal Addition

The addition of fractals does not seem to have been considered before volume 1. Later we will see fractal addition is implied in certain fractal relations between particles. We will study fractal addition in the case of Cosmos spaces.

There are several definitions of dimensions for Cosmos spaces:

---

[77] For Cosmos space of dimension r there is a subset of r dimensions in its dimension array for its space-time.

[8] For 2nd Kind HyperCosmos spaces r = n – 3.

1. We have assigned an even integer dimension r to each Cosmos space.
2. Each HyperCosmos space of dimension r has a dimension $D(2^{r+4}) = r + 4$. The number of elements in its dimension array is $d_{dr} = 2^{r+4}$.
3. Each 2nd Kind HyperCosmos space of dimension r has a dimension $D(2^{r+3}) = r + 3$. The number of elements in its dimension array is $d_{dr} = 2^{r+3}$.
4. Each HyperCosmos space and 2nd Kind HyperCosmos space of dimension r has a space-time dimension r.

We define the fractal dimension of a HyperCosmos space of dimension r + 4 with

$$D(2^{r+4}) = \log(2^{r+4})/\log(2) = \log_2(2^{r+4}) = r + 4 \quad \text{for HyperCosmos Spaces} \quad (2\text{-A.1})$$

and for 2nd Kind HyperCosmos spaces as

$$D(2^{r+3}) = \log(2^{r+3})/\log(2) = \log_2(2^{r+3}) = r + 3 \quad \text{for 2nd Kind HyperCosmos Spaces}$$

See Figs. 2.1, 2.2, and 2.3. In general we define

$$D(2^n) = \log(2^n)/\log(2) = \log_2(2^n) = n \quad (2\text{-A.2})$$

and an inverse

$$D^{-1}(n) = 2^n$$
$$D^{-1}(D(2^n)) = 2^n$$
$$D(D^{-1}(n)) = n$$

We identify the dimension array sizes as corresponding to fractal dimensions. (They are also the lengths of piecewise linear line segments in the construction of the Cosmos fractal curve.)[9]

If we "add" two spaces of dimensions r and r′, the combined space has the dimension array size:

HyperCosmos Space: $d_{dr}\, d_{dr'} = 2^{r+4} * 2^{r'+4} = 2^{r+r'+8} = 16 d_{d(r+r')}$ (2-A.3)

2nd Kind HyperCosmos Space: $d_{dr2}\, d_{dr'2} = 2^{r+3} * 2^{r'+3} = 2^{r+r'+6} = 8 d_{d(r+r')2}$

The corresponding Hausdorff dimension "additions" give the combined dimensions as[10]

$$D(2^{r+4}) \text{ “+” } D(2^{r'+4}) = D(2^{r+r'+8}) \quad (2\text{-A.4})$$

and

$$D(2^{r+3}) \text{ “+” } D(2^{r'+3}) = D(2^{r+r'+6})$$

---

99 These values change to 2^{r+7} and 2^{r+6} respectively when strata and fermion gambols are introduced. The results are similar with these choices.

10 We have chosen to use base 2 in our study of dimensions because of its appearance in Cosmos spaces. Other choices of base are possible.

It is clear that dimension addition adds the powers of 2. We can interpret the "addition" in eq. 1.4 with blocks of dimension arrays as:

Dimension array *sizes*:

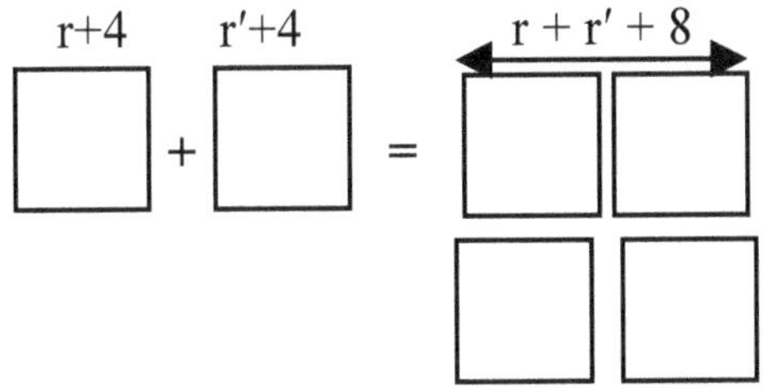

2-A.4 Fractal Subtraction

Fractal subtraction causes removal of parts of a fractal. Again we view it in the context of the dimension arrays of Cosmos spaces.

$$D(2^{r+4}) \text{ "–" } D(2^{r'+4}) = D(2^{r-r'}) \qquad (2\text{-A.5})$$
$$D(2^{r+3}) \text{ "–" } D(2^{r'+3}) = D(2^{r-r'})$$

based on

$$2^{r+4}2^{-(r'+4)} = 2^{r-r'} \qquad (2\text{-A.6})$$

$$2^{r+3}2^{-(r'+3)} = 2^{r-r'}$$

It is clear that dimension subtraction subtracts the exponents of 2. Subtraction may be viewed in terms of dimension arrays as

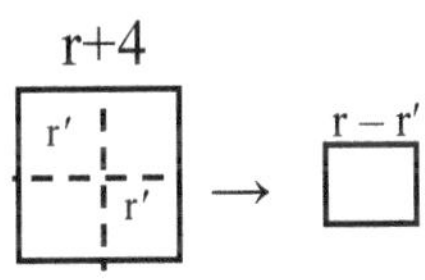

The interior r′ parts are excised by the subtraction. Fractal subtraction assumes $r > r'$.

2-A.5 Fractal Multiplication

We will define commutative fractal multiplication[11] of two fractals to yield the product of the fractal dimensions. An initial form of fractal multiplication can be implemented within the framework of Cosmos spaces:

$$D = D(2^{r+4}) \text{ "•" } D(2^{r'+4}) = (r + 4)(r' + 4) \qquad (2\text{-A.7})$$

For the multiplication of Cosmos dimension array fractals based on

$$2^{r+4} \text{ "•" } 2^{r'+4} = (2^{r+4})^{r'+4} = 2^{(r+4)(r'+4)} \qquad (2\text{-A.8})$$

[11] We have replaced the non-commutative definition of fractal multiplication of *Fractal Cosmos* with this commutative definition.

with the result

$$D = \log_2(2^{(r+4)(r'+4)}) = (r' + 4)\log_2(2^{(r+4)}) = (r + 4)(r' + 4) \quad (2\text{-A.9})$$

For 2nd Kind HyperCosmos dimension arrays

$$D = \log_2(2^{(r+3)(r'+3)}) = (r'+3)\log_2(2^{(r+3)}) = (r'+3)(r+3) \quad (2\text{-A.10})$$

Fractal multiplication for Cosmos spaces takes every element of a dimension array and expands it (through multiplication) by the dimension array of the other multiplicand.

In general[12]

$$D = D(2^n) \text{ "•" } D(2^m) = mn \quad (2\text{-A.7}')$$

2-A.6 Fractal Division

Fractal division[13] can be implemented within the framework of Cosmos spaces:

$$D = D(2^{r+4}) \text{ "/" } D(2^{r'+4}) \quad (2\text{-A.11})$$

based on

$$2^{r+4} \text{ "/" } 2^{(r'+4)} = 2^{(r+4)/(r'+4)} \quad (2\text{-A.12})$$

Then

$$D = \log_2(2^{(r+4)/(r'+4)}) = \log_2(2^{(r+4)})/(r' + 4) = (r+4)/(r' + 4) \quad (2\text{-A.13})$$

In general[14]

$$D = D(2^m) \text{ "/" } D(2^n) = m/n \quad (2\text{-A.13}')$$

Similarly for 2nd Kind HyperCosmos dimension arrays:

$$D = (r+3)/(r' + 3) \quad (2\text{-A.14})$$

Fractal division of Cosmos spaces requires factoring within the r dimension array extracting r′ dimension array factors. Thus $r > r'$ is required or the result is a non-integer fractal dimension. Eqs. 2-A.13 and 2-A.14 explicitly show the factors.

2-A.7 Extension to Real Fractals

The addition of fractals in general combines the unit squares of two fractals in a manner similar to the illustration in section 2-A.3.

The subtraction of fractals in general separates the unit squares of the two fractals in a manner similar to the illustration in section 2-A.4. The terms in the fractal constructions must be separated to obtain the result. This presumes an extraction method (up to a multiple of the construction numbers) supporting their removal.

---

[12] This expression may be generalized to powers greater than 2.

[13] We have replaced the definition of fractal division of *Fractal Cosmos* with this definition which is consistent with commutative fractal multiplication as defined above.

[14] This expression may be generalized to powers greater than 2.

The multiplication of fractals in general combines the unit squares of two fractals in a manner similar to the description in section 2-A.5. The terms in one multiplicand of its fractal construction must be "multiplied" by the terms of the construction of the second multiplicand. The result is a combined fractal with terms being fractals of the second multiplicand.

The division of fractals in general decomposes the unit square of one fractal piecewise linear curve by piecewise linear curve extraction of the divisor fractal. The terms in the dividend in its fractal construction are divided by the terms of the construction of the divisor.

3. Quantum Fractal Curve

We have created a deterministic fractal Cosmos Curve in chapter 2. We now consider the possibility that the Cosmos Curve is quantum, and generated step by step probabilistically. The amplitude for the quantum fractal generation process can be symbolized with

$$A = \prod_{\substack{n=1 \\ \text{steps}}}^{\infty} A(n, \theta_n, \varphi_n) \tag{3.1}$$

based on eq. 2.18′ where[15]

$$A(n, \theta_n, \varphi_n) = e^{i\varphi_n}\cos(\theta_n)\psi_{\text{HyperCosmos, n=r+4}} + e^{-i\varphi_n}\sin(\theta_n)\psi_{\text{2nd Kind HyperCosmos, n=r+3}} \tag{3.2}$$

with the wave functions determined by solutions of the Master Equation if one chooses to embed the fractal curve in the two dimension space of the Master Equation. *Eq. 3.1 defines an amplitude A for a quantum Cosmos Space field for each step in the construction of the Cosmos Fractal Curve. The product over n of these amplitudes is the amplitude for the creation of the Quantum Cosmos Fractal Curve set of quantum Cosmos spaces.*

The probability of this Quantum Cosmos Fractal Curve curve is

$$P = |A|^2 \tag{3.3}$$

With this probabilistic Quantum Cosmos Fractal Curve, we have a new degree of freedom in Cosmos Theory. We can now consider variations in the definition of Cosmos Theory.

The above definition of the Quantum Cosmos Fractal Curve could be more deeply based on Cosmos fields quantization, *and also* on a direct quantization of the piecewise linear curves used at each step in the construction.

3.1 Quantum Step Fractal Curve Construction

For step length quantization of the *Step* Quantum Cosmos Fractal Curve, we assign an amplitude $A(l_n)$ for the piecewise linear curve length where l is its length. (The classical limit of the nth length is 2^n.) Then the joint amplitude for the *Step* Quantum Cosmos Fractal Curve is

$$A(l) = \prod_{\substack{n=1 \\ \text{steps}}}^{\infty} A(l_n) \tag{3.4}$$

Combining eqs. 3.1 and 3.4 gives the joint amplitude for quantization of step lengths and Cosmos fields

---

[15] The amplitude A may also be formulated using eqs. 2.13 and 2.18′.

$$A_{tot} = \prod_n A_n(l)\, A(n, \theta_n, \varphi_n) \quad (3.5)$$

3.1.1 Example of Quantum Step Fractal Curve

An example model of step quantization uses a Gaussian amplitude

$$A(l_n) = N_n \exp(-(l_n - 2^n)^2)/n^2) \quad (3.6)$$

We assume

$$l_n - 2^n = \varepsilon \quad (3.7)$$

where ε is a constant for all n. Then the total amplitude is

$$A(l) = N \prod_{n=1}^{\infty} A(l_n) = N \exp(-\sum_{n=1}^{\infty} \varepsilon^2/n^2) = N \exp(-\varepsilon^2\pi^2/6) = N \exp(-\varepsilon^2 \zeta(2)) \quad (3.8)$$

using the *Basel* problem solution found by Euler where $\zeta(2)$ is a Riemann zeta function. If $|A(l)|^2$ is suitably normalized it gives the probability of a quantum fractal curve for a constant deviation ε from the "classical" value for all steps n.

$$N^{-2} = \int_{-\infty}^{\infty} d\varepsilon \exp(-\varepsilon^2 \zeta(2)) = (\pi/\zeta(2))^{1/2} = (6/\pi)^{1/2} \quad (3.9)$$

where $\zeta(2) = \pi^2/6$ and

$$N = \prod_{n=1}^{\infty} N_n = (\pi/6)^{1/4} = 0.850 \quad (3.10)$$

Probability of fractal with ε piecewise curve discrepancy $= (\zeta(2)/\pi)^{1/2} \exp(-2\varepsilon^2 \zeta(2))$ (3.11)

The probability for the Hilbert fractal curve (and the Cosmos Fractal Curve) where ε = 0 is

$$(\zeta(2)/\pi)^{1/2}\exp(-2\varepsilon^2 \zeta(2))|_{\varepsilon=0} = (\zeta(2)/\pi)^{1/2} = 0.7236 \quad (3.12)$$

This probability, obtained in a Gaussian model, is encouragingly large.

Note $N^4 = \pi/6 = 0.5236$ is approximately equal to the calculated value 0.538 for the fractal Hausdorff dimension of the Feigenbaum Attractor.

3.2 Importance of Quantum Fractal Curves

The importance of a Quantum Cosmos Fractal Curve, and the attendant Quantum considerations, lies in the resolution it provides of the question of the origin of the series of "events" for the generation of the Master Equation and related quantities from a mathematical viewpoint. One might attribute the progression of events to the actions of a Prime Mover. However a Quantum formalism can make the progression of events probabilistic without a need for an external source. The origin of the Quantum amplitudes still remains an issue. But it is a less demanding choice than attributing the progression of events to a Prime Mover executing a sequence of *decisions.*

The determination of the Quantum probability amplitudes appears to be an open Physics question that requires further investigation. It appears that the answer might be in the development of an enveloping Quantum formalism for the complete progression of events in the formation of Cosmos Theory.

3.3 Fractal Curve Probabilities

At each step in the fractal curve construction there is a corresponding Cosmos space. Consequently we can associate each step's probability amplitude defined above with a Cosmos space. Then the Cosmos space spectrums of Figs. 2.1 – 2.3 can be viewed as classical – not quantum.

We now introduce a Quantum Cosmos with the spaces being combinations of states defined by eq. 3.3. Its form for both HyperCosmos and 2nd Kind HyperCosmos spaces is

$$S(\text{HyperCosmos}, r, n, \theta_n, \varphi_n) = e^{i\varphi_n}\cos(\theta_n)S_{\text{HyperCosmos},\ n=n'=r+4} + e^{-i\varphi_n}\sin(\theta_n)S_{\text{2nd Kind HyperCosmos},\ n\pm1=n'=r+3\pm2} \qquad (3.11)$$

and

$$S(2^{nd}\ \text{Kind HyperCosmos}, r, n, \theta_n, \varphi_n) = e^{-i\varphi_n}\sin(\theta_n)S_{\text{HyperCosmos},\ n=n'=r+4} + e^{i\varphi_n}\cos(\theta_n)S_{\text{2nd Kind HyperCosmos},\ n\pm1=n'=r+3\pm2} \qquad (3.12)$$

where we specify spaces by their dimension r and the terms S… represent wave functions for dimension arrays. Eq. 3.11 applies to HyperCosmos spaces and Limos spaces. Eq. 3.12 applies to 2nd Kind HyperCosmos spaces. When $e^{i\varphi_n}\cos(\theta_n) = 1$ in eq. 3.2 we obtain the HyperCosmos spaces and Limos spaces of Figs. 1.1 and 3.12. When $e^{i\varphi_n}\sin(\theta_n) = 1$ in eq. 3.2 we obtain the 2nd Kind HyperCosmos spaces of Figs. 1.3.

We use these probabilistic spaces in the definition of universes later.

3.4 Quantum Cosmos Spaces Spectrum

We have previously based the spectrum of HyperCosmos spaces and 2nd Kind HyperCosmos spaces on totally antisymmetric tensors, and on the creation and annihilation operators of fermion wave functions. Recently we further based these spaces on the solutions of the Master Equation.

We now introduce a quantum *probabilistic* form of the Cosmos spaces spectrum. We base this form on a map between Cosmos Fractal Curve steps and Cosmos spaces developed in chapter 2.

Each space in Figs. 2.1 and 2.2 now has a probabilistic form which appears in Fig. 3.1. The 2nd Kind HyperCosmos spaces are embodied in the probabilistic entries in Fig. 3.1. We define

$$\Psi(r, \theta_r) = e^{i\varphi_r}\cos(\theta_r)\psi_H(r) + e^{-i\varphi_r}\sin(\theta_r)\psi_{H2}(r) \qquad (3.13)$$

where $\psi_H(r)$ is a wave function for the dimension array with 2^{r+4} of a HyperCosmos space, and where $\psi_{H2}(r)$ is a wave function for the dimension array with 2^{r+3} of a 2nd Kind HyperCosmos space.

These probabilistic spaces will be used later in the creation of universes.

3.5 Quantum Field Theory Formulation for the Two plus One Dimension Space

Having defined a time coordinate[16] earlier in chapter 1 we now develop a Quantum Field Theory[17] that leads to the Master Equation. The Lagrangian density is

$$\mathcal{L} = ½(\partial^\mu \varphi(z)\partial_\mu\varphi(z)/m \;-\; V(z)\varphi(z)\varphi(z)) \qquad (3.14)$$

where z is the 3-vector (t, x, y), $\partial^\mu \varphi(z)$ is the covariant derivative, a (non-mass) constant m is introduced for use later, and V(z) is a scaling potential

$$V(z) = \lambda/[m^2(x^2 + y^2)] \qquad (3.15)$$

with λ a constant, From it we derive the Master Equation seen later in chapter 4:

$$[(\partial^2/\partial x^2 + \partial^2/\partial y^2)/m + \lambda/m^2r^2 + C)]\psi = 0 \qquad (3.16)$$

where $r = (x^2 + y^2)^{½}$ after letting

$$\varphi = e^{iEt}\psi(x, y) \qquad (3.17)$$

with C – $-E^2$.

In chapter 5 we introduce quantum fields and develop a probabilistic approach to universe creation.

---

[16] A time coordinate is defined in this space as the distance along the the fractal curve. We map the curve to a time coordinate represented by a continuous "straight line." The time coordinate has an infinite range.

[17] Quantum Field Theory a more fundamental theory and implies Quantum Mechanics. Thus it is the preferred quantum formulation for Cosmos Theory. W. Heitler, *The Quantim Theory of Radiation* (Oxford University Press, Oxford, UK, 1954) provides a detailed derivation of the Uncertainty Conditions from Quantum Field Theory.

THE HYPERCOSMOS SPACES SPECTRUM

Blaha Space Number $N = o_s$	Cayley-Dickson Number	Cayley Number Cn d_c	Space-time-Dimension r	Wave Function $\Psi(r, \theta_r)$
0	10	1024	18	$\Psi(18, \theta_{18})$
1	9	512	16	$\Psi(16, \theta_{16})$
2	8	256	14	$\Psi(14, \theta_{14})$
3	7	128	12	$\Psi(12, \theta_{12})$
4	6	64	10	$\Psi(10, \theta_{10})$
5	5	32	8	$\Psi(8, \theta_8)$
6	4	16	6	$\Psi(6, \theta_6)$
7	**3**	**8**	**4**	$\Psi(4, \theta_4)$
8	2	4	2	$\Psi(2, \theta_2)$
9	1	2	0	$\Psi(0, \theta_0)$
10	0	1	-2	$\Psi(-2, \theta_{-2})$
11	-1	½	-4	$\Psi(-4, \theta_{-4})$
12	-2	¼	-6	$\Psi(-6, \theta_{-6})$
13	-3	1/8	-8	$\Psi(-8, \theta_{-8})$
14	-4	1/16	-10	$\Psi(-10, \theta_{-10})$

...

Figure 3.1. The Probabilistic Cosmos space spectrum. Note: negative r values correspond to overconstrained spaces. We discuss this issue in an earlier book.

4. Genesis of the Master Equation[18]

The Cosmos Fractal curve has two significant features: a sequence of piecewise linear segments of length 2^n for construction step n, and the compression of these sequences to fit within a unit square. In the $n \to \infty$ limit the sequence becomes a space occupying fractal.

The sequence of piecewise linear segment lengths is 2^n. We can define a function that encapsulates this combined series of lengths in the form

$$\psi(u) = \cos(\pi \log_2(u)) \tag{4.1}$$

where u is a parameter. The zeros of $\psi(u)$ are at

$$u_n = 2^n \tag{4.2}$$

for integer n. The form of this solution and its zeros are scale independent – as it should be since no scale is present for the fractal construction.

We have seen in earlier books that eq. 4.1 is a special case of the type of the solutions of the Master Equation. The Master Equation is based on a scaling $1/r^2$ potential. The equation is a second order differential equation similar to the three dimension Schrödinger equation in a three dimension space. Its derivation is performed in Quantum Field Theory in chapter 3.

The fractal Cosmos Curve has fractal dimension 2. It generates a unit square. This unit square may be scaled to an infinite two dimension space to which we added a time coordinate based on the distance along the Cosmic Curve.[19] The Master Equation is defined within the framework of this space.

It is a Schrödinger-like equation with a $1/r^2$ potential. The equation scales and allows us to avoid the introduction of a mass scale. The scaling $1/r^2$ potential has the advantages:

1. No mass scale dependence.
2. A spectrum of powers of a multiplier times an overall constant.
3. A constant that can be scaled out of the Master Equation by scaling the radial coordinate.
4. The capability to scale any energy eigenvalue to one.

---

[18] This chapter is an enhanced version of the corresponding chapter in Blaha (2025g).

[19] Item 7 in chapter 1.

5. The generality of the scaling $1/r^2$ Master Equation solution is supported by its solution also being equivalent to a harmonic oscillator-like solution and also a Free Field solution for small values of the radial parameter.

Perhaps the most significant feature of the $1/r^2$ potential is that it generates a set of eigenvalues in the form of a sequence of powers of a constant. This feature corresponds to the power series of the Cosmos spaces dimension arrays. See Fig. 2.6 where they are listed.

4.1 Master Equation Forms and Solutions[20]

The scaling potential provides the simplest guarantee of an energy spectrum that is a sequence of powers of a constant.[21] The Master Equation is:[22]

$$[(\partial^2/\partial x^2 + \partial^2/\partial y^2)/m + \lambda/m^2r^2 + C)]\psi = 0 \tag{4.3}$$

where λ is a coupling constant, m is a constant, and $r = (x^2 + y^2)^{½}$. In polar coordinates

$$x = u\cos(\varphi)$$
$$y = u\sin(\varphi)$$

we find

$$(\partial^2/\partial u^2 + 1/u\,\partial/\partial u + \partial^2/\partial\varphi^2 + \lambda/mu^2 + mC)\psi = 0 \tag{4.4}$$
$$(\partial^2/\partial\varphi^2)\psi = -k^2\psi$$

based on the solution

$$\psi = \psi(u)\Phi(\varphi) \tag{4.5}$$

where

$$\Phi(\varphi) = \exp(ik\varphi) \tag{4.6}$$

The quantum number k must be a whole integer to avoid irregularities in $\psi(r)$. We set $\lambda = k^2 + n_\varphi^2$. Substituting and separating factors in eq. 4.4 we find

$$(\partial^2/\partial u^2 + 1/u\,\partial/\partial u + (\lambda - k^2)/(mu^2) + mC)\psi(u) = 0$$
$$(\partial^2/\partial u^2 + 1/u\,\partial/\partial u + \lambda_{n\varphi}/u^2 + mC)\psi(u) = 0 \tag{4.7}$$

where[23] we use the constant $(-mc_{n\varphi})^{½}$

$$\lambda_{n\varphi} = n_\varphi^2/m \tag{4.8}$$
$$E_{n\varphi n} = (-mC_{n\varphi n})^{½} = (-mc_{n\varphi})^{½}\exp(-n\pi/(\lambda_{n\varphi} - ¼)^{½}) \tag{4.9}$$

---

[20] The discussions of these sections appear in more detail in Blaha (2025c).

[21] This equation may be viewed within the context of conformal symmetry due to its local scale invariant potential.

[22] This differential equation might be generalized to a fractal differential equation. However the coordinates are c-numbers. So a fractal extension seems unwarranted.

[23] The eigenvalue spectrum results from the imposition of orthogonality condition on the solution function near u = 0 according to von Neumann. The value of $-mC_{n\varphi n}$ can be taken to be positive since its sign can be changed by Wick rotations of the time variable: $t \rightarrow it$.

Or, upon letting[24] $n \to -n$, we have

$$C_{n\varphi n} = c_{n\varphi} \exp(2\pi n/(\lambda_{n\varphi} - ¼)^{½}) \quad (4.9a)$$

where n is a radial quantum number (related to r), n_φ is the angular momentum quantum number, and $c_{n\varphi}$ is an arbitrary constant.[25] The growth factor (multiplier) for each power series sequence is

$$g_{n\varphi} = \exp(2\pi/(\lambda_{n\varphi} - ¼)^{½}) \quad (4.10)$$

4.2 The Solutions of the Scaling Master Equation

The solution of eq. 4.7 of the Master Equation can be transformed to a function w(r) that makes the equation Schrödinger-like:

$$\psi(u) = u^{-½}\, w(u) \quad (4.11)$$

and obtain

$$(\partial^2/\partial u^2 + \lambda_{n\varphi}/(mu^2) + mC)w(u) = 0 \quad (4.12)$$

where

$$\lambda_{n\varphi} = n_\varphi^{\,2}/m \quad (4.13)$$

and n_φ is a positive or negative integer. We now consider the solutions of eq. 4.12 for small r. The complete solution is a sum of Bessel functions.[26]

The solution for w(u) for small u is[27]

$$w(u) \approx Br^{½} \cos((\lambda_{n\varphi} - ¼)^{½}\ln((-mC_{n\varphi n})^{½}u) + b) \quad (4.14)$$

implying

$$\psi(u) \approx B\cos((\lambda_{n\varphi} - ¼)^{½}\ln((-mC_{n\varphi n})^{½}u) + b) \quad (4.15)$$

where B and b are constants and ln() is the natural logarithm $\ln_2(u)$. The solution eq4.15 may be expressed as

$$\psi(u) \approx B\cos((\lambda_{n\varphi} - ¼)^{½}\ln(u) + b_{n\varphi n}) \quad (4.15a)$$

where

$$b_{n\varphi n}) = b + ½ \ln(-mC_{n\varphi n})$$

Eq. 4.15a has a form similar to eq. 4.1 showing the close similarity of the zeros of the solutions of eq. 4.14 to the Cosmos spaces of the Cosmos Fractal Curve.

4.3 Other Related Forms of the Master Equation

The Master Equation has other familiar forms when the parameter u is suitably transformed. It has a harmonic oscillator-like form and also a free wave function form.

---

[24] The sign of n can be positive or negative due to the scaling property of the Master Equation.
[25] The constant $c_{n\varphi}$ can be positive or negative. Its sign can be changed by Wick rotations of the time variable: $t \to it$.
[26] Since we can scale the r coordinate due to the scale covariant $1/r^2$ potential we can take the small r solution to be extended indefinitely.
[27] K. M. Case, Phys. Rev. **80**, 797 (1950). eq. 16.

If we define a variable s with

$$s = \ln u \tag{4.16}$$

then

$$(\partial^2/\partial s^2 - \partial/\partial s + \lambda_{n\varphi}/m - mCu^2)w(r) = 0 \tag{4.17}$$

Letting

$$w(u) = e^{s/2}\, v(s) = u^{½}\, v(s) \tag{4.18}$$

where, from eq. 4.11, we obtain

$$v(s) = u^{-½}\, w(u) = \psi(u) \tag{4.19}$$

Then we find

$$(\partial^2/\partial s^2 + (\lambda_{n\varphi}/m - ¼) \;\; - mCu^2)v(s) = 0 \tag{4.20}$$

or

$$(\partial^2/\partial s^2 + (\lambda_{n\varphi}/m - ¼) \;\; - mCe^{2s})v(s) = 0 \tag{4.21}$$

The mCe^{2s} term is the potential after the transformation to a harmonic oscillator-like form. We now connect the v(s) solution[28] for small u and thus large negative s to an approximate ψ(u) solution for small u given in eq. 4.15. We approximate eq. 4.24 with

$$(\partial^2/\partial s^2 + \lambda_{n\varphi}/m - ¼\,)v(s) = 0 \tag{4.22}$$

Eq. 4.25 is a free equation for v(s). The generality of the Master Equation solution is supported by its solution being equivalent to a Free Field solution (eq. 4.27 below) for small values of the radial parameter u.

A relevant form of the solution for comparison with eq. 4.15 is (using similar constants)

$$v(s) \approx B\cos((\lambda_{n\varphi} - ¼)^{½}\, s + b) \tag{4.23}$$

Note eq. 4.15a can be expressed as

$$\psi(u) \approx B\cos((\lambda_{n\varphi} - ¼)^{½} s + b) \tag{4.24}$$

Eq. 4.22 corresponds to a free quantum field equation. There is no direct quantization of the energy. The energy is a positive number. However we can impose a quantization based on a requirement of orthogonality of solutions (an observation due to von Neumann and Case) as we did in the scaling case. Then the energy is quantized due to the form of the solution in eqs. 4.23 and 4.24 for s. The quantized energy $E_{n\varphi n}$ is

$$E_{n\varphi n} = n\pi/(\lambda_{n\varphi} \; - ¼)^{½} \tag{4.25}$$

corresponding to the wave function

---

[28] We approximate the potential at its minimum relying on the scalability of u. Thus the small u (large negative s) neighborhood represents the potentials well.

$$\psi_{n\varphi n}(u) \approx B_n \cos((\lambda_{n\varphi} - \frac{1}{4})^{\frac{1}{2}} \ln(u) + b_n) \tag{4.25a}$$

Its form agrees with the scaling Master Equation solution of eq. 4.9:

$$C_{n\varphi n} = c_{n\varphi} \exp(2n\pi/(\lambda_{n\varphi} - \frac{1}{4})^{\frac{1}{2}}) = c_{n\varphi} \exp(2E_{n\varphi n}) \tag{4.26}$$

The constant n can be positive or negative. We choose n to be positive with n and n_φ being independent quantum numbers.

We thus find agreement between this approximate harmonic oscillator form of the Master Equation and the scaling Master Equation.

The transformation to an approximate harmonic oscillator form implies a form of stability for the Master Equation solutions. It also suggests a universality due to the generality of harmonic oscillator solutions under small perturbations for many Physical forces.

4.4 Cosmos Spaces Dimension arrays

Cosmos spaces dimension arrays have two forms: HyperCosmos Spaces and 2nd Kind HyperCosmos Spaces Dimension arrays. They are based on eq. 4.9a and the choices of the $c_{n\varphi}$ constants. They both set the multiplier factor

$$\exp(2\pi/(\lambda_{n\varphi} - \frac{1}{4})^{\frac{1}{2}}) - 2 \tag{4.27}$$

4.4.1 HyperCosmos Solutions Eigenvalues

To obtain the HyperCosmos spaces dimension arrays we then set[29]

$$n = r + 4 \tag{4.28}$$

with $c_{n\varphi} = 1$. From eqs. 4.9a and 4.27 we find the dimension r dimension array

$$C_{n\varphi n} \equiv C_r = 2^{r+4} \tag{4.29}$$

4.4.2 Second Kind HyperCosmos Solutions Eigenvalues

To obtain the 2nd Kind HyperCosmos spaces dimension arrays we set

$$n = r + 3 \tag{4.30}$$

and $c_{n\varphi} = 1$ resulting in the dimension r dimension array

$$C_{n\varphi n} = C_{r2} = 2^{r+3} \tag{4.31}$$

---

[29] We could alternately place the 4 (and 3) appearing in eq.s 4.28 and 4.29 in the $c_{n\varphi}$ factors.

4.5 Distinction between HyperCosmos spaces and 2nd Kind HyperCosmos spaces

The steps of the Cosmos Curve construction process display a feature that relates to the distinction between HyperCosmos spaces and 2nd Kind HyperCosmos spaces.[30] The feature follows from our requirement that r be an even integer.[31] There is an important *Physical* distinction between these types of spaces. HyperCosmos spaces' dimension arrays have both a Normal sector and a Dark sector while 2nd Kind HyperCosmos spaces' dimension arrays have only a Normal sector. The difference is a factor of 2 in the sizes of dimension arrays. Our UST is an example since it has an r = 4 HyperCosmos space dimension array with $2^{4+4} = 256$ components and has both a Normal and Dark sector. A *different* choice of a 2nd Kind HyperCosmos dimension array with $2^{4+3} = 128$ components would only be a Normal sector as we discussed in earlier books.

[30] A HyperCosmos space has both a Normal and a Dark sector. A 2nd Kind HyperCosmos space only has a Normal sector. As a result their dimension array sizes differ by a factor of 2.

[31] The value of r is even to prevent double counting of spaces since an r+1 dimension array is the same as an r dimension array.

Appendix 4-A. The Primeval Point - Master Equation Connection

The primeval Mathematical point leading to the Fractal unit square and to a two dimension space superficially might appear to be disjoint from the Master Equation. The Master Equation generates a progression of structure that leads fairly directly to coupling constant values and fundamental fermion masses.

At first glance the progression from Mathematical point to two dimension space does not appear to have experimental support. Thus there is a perceived conceptual "gap" between the progression and the Master Equation.

There are a number of reasons supporting the connection of the progression to the Master Equation:

1. The Master Equation solutions have zeros that can be put into a one to one correspondence with the piecewise linear line segments used in the construction of the fractal Cosmos curve that leads to the 2D space for the Master Equation. The values of $u_n = 2^n$ specify the length of the piecewise linear line segments that generate a Cosmos (and Hilbert) fractal curve. Other choices of the $g_{n\varphi}$ value generate other power sequences that can be scaled to the Hilbert sequence powers of 2^n. One can view the Master Equation as implying the fractal construction process. Thus the construction process from a point to the fractal-based unit square can be viewed as embodied in the Master Equation. The fractal construction process has an integral part in the generation of Cosmos structure.

2. The Master Equation solutions can be viewed as defining a generating function for the lengths of the piecewise linear line segments of the fractal.

3. The Master Equation has features that make it the simplest, and possibly the most general, choice of possible Cosmos generating equations:

    a. First it is a second order differential equation similar to the two dimension Schrödinger equation.
    b. It has a $1/u^2$ potential that enables the avoidance of the introduction of a mass scale.
    c. It can be transformed to a harmonic oscillator-like form with an equally spaced eigenvalue spectrum.
    d. The generality of the scaling Master Equation solution is supported by its solution also being equivalent to a Free Field solution for small values of the radial parameter.

4. The fractal curve construction supports a map between the 1D piecewise linear segments and the 2D unit square and space that preserves locality: two data points that are close to each other in a 1D line segment are close to each other after the line is folded in the fractal construction process. Locality is a very desirable feature in Physics.

5. The Master Equation, which embodies scaling, can be transformed to a harmonic oscillator-like form with a corresponding eigenvalue spectrum. This form, which characterizes small perturbative forces, provides a type of stability against disturbances from small forces.

    a. The generality of the scaling Master Equation solution is supported by its solution also being equivalent to a Free Field solution for small values of the radial parameter u.

6. We have shown from the Master Equation solution that the evolutionary sequence of the Cosmos follows. We see it encompasses the initial mathematical point, the generation of a two dimension piecewise linear unit length, the generation of the fractal Cosmos Curve, the generation of the Cosmos spaces and their dimension arrays, the unification spaces sequence, and the set of sequences of Fig. 5.2.

The Master Equation is the basis of the evolution of the Cosmos conceptually. As we have often pointed out, Cosmos evolution is instantaneous in the sense that there is no time coordinate. Therefore it is timeless! The Master Equation embodies the Cosmos and sets the stage for its universes. The unity of the progression from point to Master Equation is established.

5. Cosmos Space States and Universe Creation

Having defined the Quantum Cosmos Spaces we are now in a position to define the initial states of universes. We begin by defining Quantum states for Cosmos spaces based on the Cosmos quantization described in chapter 3.

Then we define the quantum process for the selection of a Cosmos space. We will specify a fractal curve state for the Quantum Cosmos Curve defined in chapter 3. This quantum fractal takes a point – the creation point of a universe with a certain energy through an unspecified process such as an inter-universe interaction – and generates the universe with a certain space dimension r and a space-time of dimension r.

We view the universe generation as proceeding through a transformation of the one dimension quantum fractal curve to an r space-time dimension space filling curve that becomes a filled r dimension "unit" hypercube. The transformation inverse[32] maps a one dimension fractal to a Z-order Lebesgue r dimension space filling curve (Morton space-filling curve).

The hypercube is then scaled to be an infinite r dimension space-time within which dynamics can occur.

Given the total energy (measured in its Parent universe) and its dimension array specifying its fundamental particles and interactions the universe proceeds to evolve dynamically.

This chapter develops the formalism, in some detail, for a child universe wave function within a Parent universe based on a selection of Cosmos space and positioning of a child universe within the Parent universe.

A universe is a fractal construct embodying complex dynamics.

5.1 Cosmos Space Master Equation Solution

Chapter 4 provides the solution of the Master Equation. It defines

$$\psi = \psi(u)\Phi(\varphi) \tag{4.5}$$

where

$$\Phi(\varphi) = \exp(ik\varphi) \tag{4.6}$$

The form of the complete wave function solution of the Master Equation is

$$\psi_{E,\, k,\, n\varphi,\, n}(u, t) = e^{iEt} \exp^{ik\varphi} \psi_{n\varphi n}(u) \tag{5.1}$$

The $\psi(u)$ radial factor solution is

$$\psi_{n\varphi n}(u) \approx B\cos((\lambda_{n\varphi} - 1/4)^{1/2} \ln((-mC_{n\varphi n})^{1/2} u) + b) \tag{4.15}$$

[32] A Z-order Lebesgue curve (Morton space-filling curve) can map multi-dimension matter to a one dimension curve.

where $c_{n\varphi}$, b and B are constants and

$$C_{n\varphi n} = c_{n\varphi} \exp(2\pi n/(\lambda_{n\varphi} - 1/4)^{1/2}) \tag{4.9a}$$

5.2 Cosmos Master Equation 2D Space Wave Function Expansions

The solution of the Master Equation eq. 4.15 can be used to define a wave function for the Master Equation Lagrangian in *its* two dimension space augmented with a time coordinate. Treating the universe in question as spin 0 and having a HyperCosmos space we find it has the form of a HyperCosmos space we find

$$\psi_{Hn\varphi n}(\varphi, u, t) \approx \sum_{\text{even r}} \int dk\, N_H(k, r)\{a_H(k, r)\, e^{-iE_{Hn\varphi n}t - ik\varphi}(m^{1/2}2^{r/2+2}u)^{-i\lambda'_{n\varphi}} + $$

$$+ a_H^{\dagger}(k, r)\, e^{iE_{Hn\varphi n}t + ik\varphi}(m^{1/2}2^{r/2+2}u)^{i\lambda'_{n\varphi}}\} \tag{5.2}$$

and, for use below,

$$\psi_{Hn\varphi n}(k, r, \varphi, u, t)^{+} = e^{-iE_{Hn\varphi n}t - ik\varphi}(m^{1/2}2^{r/2+2}u)^{-i\lambda'_{n\varphi}} \tag{5.2a}$$

$$\psi_{Hn\varphi n}(k, r, \varphi, u, t)^{-} = e^{iE_{Hn\varphi n}t + ik\varphi}(m^{1/2}2^{r/2+2}u)^{i\lambda'_{n\varphi}} \tag{5.2b}$$

where N(k, r) is a normalization and

$$n = r + 4 \tag{4.28}$$

$$C_{Hn\varphi n} \equiv C_{Hr} = 2^{r+4} \tag{4.29}$$

$$\lambda'_{n\varphi} = (\lambda_{n\varphi} - 1/4)^{1/2} \tag{5.3}$$

$$E_{HHn\varphi n}(r) = (-mC_{Hn\varphi n})^{1/2}$$

by examining the expansion of the cosine in eq. 4.15.

For the case of a 2nd Kind HyperCosmos space we find the wave function expansion

$$\psi_{2Hn\varphi n}(\varphi, u, t) \approx \sum_{\text{even r}} \int dk\, N_{2H}(k, r)\{a_{2H}(k, r)\, e^{-iE_{2Hn\varphi n}t - ik\varphi}(m^{1/2}2^{r/2+3/2}u)^{-i\lambda'_{n\varphi}} + $$

$$+ a_{2H}^{\dagger}(k, r)\, e^{iE_{2Hn\varphi n}t + ik\varphi}(m^{1/2}2^{r/2+3.2}u)^{i\lambda'_{n\varphi}}\} \tag{5.4}$$

and

$$\psi_{2Hn\varphi n}(k, r, \varphi, u, t)^{+} = e^{-iE_{2Hn\varphi n}t - ik\varphi}(m^{1/2}2^{r/2+3/2}u)^{-i\lambda'_{n\varphi}} \tag{5.4a}$$

$$\psi_{2Hn\varphi n}(k, r, \varphi, u, t)^{-} = e^{iE_{2Hn\varphi n}t + ik\varphi}(m^{1/2}2^{r/2+3.2}u)^{i\lambda'_{n\varphi}} \tag{5.4b}$$

with

$$n = r + 3 \tag{4.30}$$

$$C_{2Hn\varphi n} \equiv C_{2Hr} = 2^{r+3}$$

The requirement that r be even is due to the form of the Cosmos Theory space spectrum. It prevents double counting of spaces since the dimension arrays are the same for adjoining even and odd dimensions.

The above wave functions can be used to define a joint wave function like eq. 3.5:

$$\Psi_{n\varphi n}(\varphi, u, t, \theta_r) = e^{i\alpha} \cos(\theta_r)\psi_{Hn\varphi n}(\varphi, u, t) + e^{-i\alpha} \sin(\theta_r)\psi_{2Hn\varphi n}(\varphi, u, t) \quad (5.5)$$

This wave function in the two dimension Master Equation space can be used in the creation/specification of a universe. It can combine with a wave function part in the Parent space parameterized by X^μ where $\mu = 0, 1, 2, \ldots, R-1$ to define a joint wave function for a particle resident at X^μ with Cosmos space properties specified by $\Psi_{n\varphi n}(\varphi, u, t, \theta_r)$.

5.2.1 Uncertainty Principle for Spaces

An examination of the Master Equation solutions and the above wave function expansions suggests an Uncertainty Relation between space dimension and Master Equation two dimension space radial coordinate u:

$$[u, r] = i\hbar \quad (5.4)$$

5.3 Cosmos Space – Parent Universe Wave Function in R Dimensions

We now turn to define a wave function for a Parent universe in the Parent universe. We use the Master Equation (eq. 4.7) extended to an r dimension[33] boson child universe of mass-energy M in the Parent universe of dimension R:

$$(\partial/\partial X_\mu \partial/\partial X^\mu + M^2 + \partial^2/\partial u^2 + 1/u\, \partial/\partial u + \lambda_{n\varphi}/u^2 + mC)\psi(X, u) = 0 \quad (5.5)$$

The solution *could* have the separated form

$$\Psi_{n\varphi n}(X, \varphi, u, t, \theta_r) = \psi(X)\Psi_{n\varphi n}(\varphi, u, t, \theta_r) \quad (5.7)$$

However the Fourier expansions of the product lead to issues with the products of raising and lowering operators. We therefore choose a composite form similar to composite forms in books such as Blaha (2023d) and (2024j).

$$\Psi_{n\varphi n}(X, \varphi,u,t,\theta_r) = \sum_{\text{even } r} \int dk d^{R-1}q\, N(q, k, r) \cdot$$
$$\cdot\{a(q,k,r)e^{-iq\cdot X}[\cos(\theta_r)\psi_{Hn\varphi n}(k,r,\varphi,u,t)^+ + \sin(\theta_r)\psi_{2Hn\varphi n}(k,r,\varphi,u,t)^+] +$$
$$+ a^\dagger(q,k,r)e^{iq\cdot X}[\cos(\theta_r)\, \psi_{Hn\varphi n}(k,r,\varphi,u,t)^- + \sin(\theta_r)\psi_{2Hn\varphi n}(k,r,\varphi,u,t)^-]\} \quad (5.8)$$

5.3.1 Commutation Relations

The commutation relations are:

$$[a(q,k,r), a^\dagger(q',k',r')] = \delta_{rr'}\delta(k - k')\delta^{R-1}(q - q') \quad (5.9)$$

---

[33] Cosmos Theory allows spaces and space-times of any dimension. A perturbation theory in a higher dimension is fully renormalizable for any polynomial Quantum Field Theory using the author's Two Tier formalism. See Blaha (2002) for details.

5.3.2 Universe States

The Universe states can be defined as products of raising operators. A simple child universe state within the Parent universe has the definition:

$$|q',k',r'\rangle = a^\dagger(q',k',r')|0\rangle \qquad (5.10)$$

5.4 Particle in the Parent Universe Wave Function

The general equation for a particle in the Parent universe has the form.

$$(\partial/\partial x_\mu \partial/\partial x^\mu + \partial/\partial X_\mu \partial/\partial X^\mu + M^2 + \partial^2/\partial u^2 + 1/u\ \partial/\partial u + \lambda_{n\varphi}/u^2 + mC)\psi(x,X,u) = 0 \quad (5.11)$$

where x^μ are the particle coordinates. The solution *could* have the separated form

$$\Psi_{n\varphi n}(x, X, \varphi, u, t, \theta_r) = \psi_{child}(x)\psi(X)\Psi_{n\varphi n}(\varphi, u, t, \theta_r) \qquad (5.12)$$

Again, the Fourier expansion of this product leads to issues with the products of raising and lowering operators. We therefore again choose a composite form.

$$\Psi_{n\varphi n}(x,X,\varphi,u,t,\theta_r) = \sum_{\text{even } r} \int dk d^{R-1}q\, d^{r-1}p\, N(p, q, k, r) \cdot$$
$$\cdot\{a(p,q,k,r)\, e^{-ip\cdot x} e^{-iq\cdot X}[\cos(\theta_r)\psi_{Hn\varphi n}(k,r,\varphi,u,t)^+ + \sin(\theta_r)\psi_{2Hn\varphi n}(k,r,\varphi,u,t)^+] +$$
$$+ a^\dagger(p,q,k,r) e^{iq\cdot X} e^{ip\cdot x}[\cos(\theta_r)\, \psi_{Hn\varphi n}(k,r,\varphi,u,t)^- + \sin(\theta_r)\psi_{2Hn\varphi n}(k,r,\varphi,u,t)^-]\} \qquad (5.13)$$

The commutation relations are:

$$[a(p,q,k,r), a^\dagger(p',q',k',r')] = \delta_{rr'}\delta(k - k')\, \delta^{r-1}(p - p')\delta^{R-1}(q - q') \qquad (5.14)$$

The particle states within the Parent universe can be defined as products of raising operators. A simple particle state within the Parent universe has the definition:

$$|p',q',k',r'\rangle = a^\dagger(p',q',k',r')|0\rangle \qquad (5.15)$$

Thus we have a complete description of a particle in the Parent universe. In contrast, the usual boson commutator and single particle state, which suffices if one views the Cosmos space and universe as not relevant in Elementary Particle Physics due to the differing energy scales, are:

$$[a(p), a^\dagger(p')] = \delta^{r-1}(p - p') \qquad (5.16)$$

$$|p\rangle = a^\dagger(p)|0\rangle \qquad (5.17)$$

If one wishes to consider the "birth" of child universes then the above full description becomes very relevant.

The case of a particle in a child universe of the Parent universe requires a straightforward extension of the above considerations.

5.5 Interpretation of Wave Function Superposition

The probabilistic form of eq. 5.5 raises the issue of its interpretation. It specifies a "superposition" of dimension arrays with sizes 2^{r+4} and 2^{r+3}. The only reasonable interpretation appears to require the 2^{r+3} size dimension array to expand with zero entries to the size 2^{r+4}.

We consider the case of r = 4 (our universe) where the square dimension arrays can be symbolized with

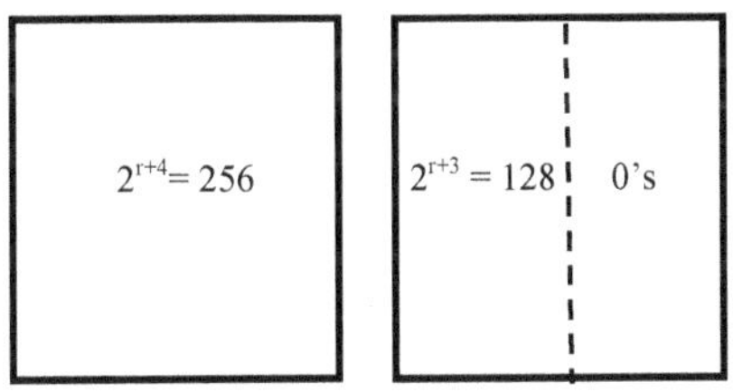

The 256 dimension array has Normal and Dark sectors of 128 dimensions. We choose the right array to have a 128 dimension Normal sector and an expansion part of 128 dimensions filled with zeros.

Each dimension array has a corresponding, but different, set of internal symmetries and interactions. We take the maps to the fermions to coincide for the Normal fermions and the 2^{r+3} size dimension array to not have Dark fermions.

Based on this structure we find, *if the angle θ_r in eq. 5.5 is infinitesimally close to $\pi/2$:*

1. The Normal fermions are predominantly 2^{r+3} dimension array fermions corresponding to the wave function $\psi_{2Hn\varphi n}(\varphi, u, t)$ and enjoy the corresponding UST interactions amongst each other.

2. The Dark fermions of the 2^{r+4} size dimension array are suppressed by a factor of $\cos(\theta_r)$ in their UST Connection Group interactions with the Normal 2^{r+4} size dimension array fermions. Thus the interaction between Normal and Dark fermions is much surpressed by $\cos(\theta_r) \approx 0$.

3. Dark fermions interact with each other with Dark interactions. The Dark interactions do not support Dark fermion – Normal fermion interactions.

4. Therefore the Dark fermions (matter) are essentially independent of the Normal fermions as far as UST (Standard Model) interactions are concerned.

5. The only remaining interaction of note is Gravitation. We expect gravitation is the only significant interaction between Normal and Dark matter.

The above view of the implications of eq, 5.5 appear to be borne out by experiment. Therefore we are essentially in a universe with a 2nd *Kind HyperCosmos space having a* $2^{r+3} = 128$ *size dimension array based on an angle* θ_r. ***slightly*** *less than* $\pi/2$.

5.6 Inverse Transformation to a Z-order Lebesgue r dimension space filling curve

When a universe is created with r dimensions it is reasonable to ask the origin of the dimensions. We view a universe as originating at a point in a Parent universe with zero dimensions, and growing in a manner similar to the Cosmos Fractal curve the difference being the dimension – the Cosmos Fractal Curve grows by amalgamating piecewise linear line segments to a two dimension space for the Master Equation. A universe can be viewed as growing from a point through a set of piecewise linear line segments to an r dimension space.

One way to realize this form of growth is the use of an inverse[34] map from a set of piecewise linear curves to a Z-order Lebesgue r dimension space filling curve (Morton space-filling curve).

The mechanism of growth can take the form of dissecting the piecewise linear curves into sequences of r bits with the accumulation of corresponding bits (each set of accumulated bits is now fractal) into r fractal pieces to form a fractal r-dimension hypercube and thence an r dimension space – the space of the universe under construction. The hypercube can be scaled to be an infinite r dimension space-time within which dynamics can occur. An r dimension universe is generated dynamically.

Given the total energy of the universe under construction (measured in its Parent universe) and its dimension array (specified with a probability amplitude – described in sections 5.3 and 5.4) specifying its fundamental particles and interactions, the universe evolves dynamically. *A universe is a fractal construct embodying complex dynamics.*

[34] A Z-order Lebesgue curve (Morton space-filling curve) can map multi-dimension matter to a one dimension curve.

6. Scaling Planckian Distribution

Some years ago we developed an apparently successful gambol theory[35] of fundamental fermions and of hadrons based on a gambol substructure with gambols having a mass that is a fraction of the particle's mass. Recently we developed a new version of the Gambol Theory[36] in which each particle has a substructure consisting of 844242 gambols that were of a new fermion type, Gambol Fermions.

In this chapter we extend the new Gambol Theory to have a new *scaling* form of Planckian distribution of gambol energies. In the following chapter we successfully apply this theory to lepton-nucleon deep inelastic scattering.

6.1 Origin of the Scaling Planckian

We now consider a modified Planckian distribution. We will see the modified distribution has the scaling property characteristic of fractal curves. There is some support for believing the scaling property is present in fundamental fermion internals. For example, the electromagnetic vacuum polarization $\Gamma(p)$ has been shown to have scaling in extensive calculations to all orders in perturbation theory:[37]

$$\Gamma_\mu(p) = f(\gamma_\mu + 2g\gamma\cdot pp_\mu/p^2)(p/\Lambda)^{2g} \tag{6.1}$$

$$\Gamma(p) = (p/\Lambda)^{2g} \tag{6.2}$$

where g is negative, f is a constant and Λ is a cutoff.

Based on this behavior and the nature on the Master Equation we have developed a scaling Planckian distribution.

6.2 A Gambol-based Derivation

We begin by noting that Einstein's results, while good for macroscopic black bodies, appear to need modification when applied to the interior of fundamental fermions. In particular, Einstein's distribution $u(E) \to 0$ as $E \to 0$..

When we examine $\Gamma(p) = (p/\Lambda)^{2g}$ for *negative* g we find it becomes zero at $p = \infty$. This view is consistent with the finiteness of the Johnson-Baker-Willey form of QED.

Therefore we propose to modify the states in the above derivation for occupation in thermal equilibrium to the form:

---

[35] Blaha (2024a).

[36] Blaha (2025g).

[37] Stephen Blaha, Phys. Rev. **D9**, 2246 (1974). See Blaha (2025b) for a detailed account of Γ and the calculation of the Fine Structure Constant α to 13 places to all orders in perturbation theory. Other calculations of α typically calculate it based on questionable extrapolations of 4th order perturbation theory.

$$\exp(-E_1/kT) \rightarrow 1/(\exp(E_1/kT) - 1) \tag{6.3}$$

and

$$\exp(-E_2/kT) \rightarrow 1/(\exp(E_2/kT) - 1) \tag{6.4}$$

The consequences of this choice are:

1. Both occupation factors become zero at $E_i = \infty$ corresponding to $\Gamma(p)$ above.

2. Since $E_2 \geq E_1$ eqs. 6.3 and 6.4 are consistent.

In the standard Planck distribution form we must make the replacement:[38]

$$1/\exp((E_2 - E_1)/kT) - 1) \rightarrow (\exp(E_1/kT) - 1)/\{(\exp(E_2/kT) - 1)\} \tag{6.5}$$
$$= (\exp(-(E_2 - E_1)/kT) - \exp(-E_2/kT)/(1 - \exp(-E_2/kT))$$

which becomes

$$= \exp(-(E_2 - E_1)/kT)$$
$$= \exp(-E/kT) \tag{6.6}$$

for large E_2.

Then we obtain the *Scaling Distribution*:

$$u(E) = (S_{12}/I_{12})/(\exp((E_2 - E_1)/kT) - 1) \tag{6.7}$$
$$= (S_{12}/I_{12})\exp(-E/kT)$$
$$= (S_{12}/I_{12})(e^E)^{-1/kT}$$

Now we see $u(E) \rightarrow 0$ as $E \rightarrow \infty$. The form of u(E) in eq. 6.7 displays scaling in E in a manner expected of fractal curves. As one scales u(E) downward, one sees a repetitive scale behavior – again a fractal feature.

We conclude the dynamic interiors of fundamental fermions and hadrons are fractal. Since a fundamental fermion contains 844242 gambol fermions, a statistical representation of their internal dynamics is warranted.

Thus we have found the fractal form of the fermion internal energy distribution. It can now be used to do gambol calculations similar to those in

Cosmos Theory: The Sub-Particle Gambol Model (Pingree Hill Publishing, Auburn, NH, 2023).
and
Cosmos-Universe-Particle-Gambol Theory (Pingree Hill Publishing, Auburn, NH, 2024).

6.3 Normalization of the Scaling Planckian Distribution

We note the number of nodes between E and E + dE $= 8\pi E^2 dE$. As a result

$$S_{12}/I_{12} = 8\pi E^3 \tag{6.8}$$

and so

---

[38] See Blaha (2025g) for a more complete discussion.

$$u(E) = 8\pi E^3 \exp(-E/kT) \tag{6.9}$$

resulting in

$$\int_0^\infty dE\, u(E) = 48\pi/(kT)^4 \tag{6.10}$$

The Scaling Planckian Distribution can be changed into a probability distribution U(E) by normalizing using eq. 6.10:

$$U(E) = (E/kT)^3/(6kT) \exp(-E/kT) \tag{6.11}$$

Note this distribution is zero at E = 0.

7. Application of Scaling Planckian to Deep Inelastic Electron-Proton Scattering

This chapter applies the scaling Planckian distribution of chapter 6 to calculate the Deep Inelastic Electron-Proton e-p structure function $F(\omega) = \nu W_2(\nu, q^2)$.[39] The author's previous calculation seems quite successful. See Fig. 7.1.

Fig. 7.3 displays $F(\omega)$ based on the Scaling Planckian distribution derived in chapter 6. The match with the experimental data is excellent. Thus there is support for scaling and thus fractal dynamics in the interior of protons (and other hadrons and fundamental fermions.) The scaling distribution is analogous to the fractal distribution of the curve of shorelines. There the fractality follows from the dynamics of wave action on the shorelines. In the case of fundamental particles, the internal dynamics apparently leads to fractal scaling Planckian distributions.

7.1 Previous Calculation of F(ω) using a Planckian Distribution

The Planckian probability distribution used *previously* for e-p scattering was:

$$F(\omega)= \Sigma_i Q_i^2 F_q(\omega) = (4/9 + 4/9 + 1/9) F_q(\omega) = F(\omega) \tag{7.1}$$

for the three quarks in a proton. The deep inelastic gambol Planckian distribution is

$$U_g(\varepsilon(s)) = 15 N (K/\pi M)^4 \varepsilon^3/(e^{\varepsilon K/M} - 1) \tag{7.2}$$

where N is a normalization factor for the photon. The gambol ω distribution for $F(\omega)$ was

$$F(\omega) = U_g(\varepsilon(2M\omega)) \tag{7.3}$$

with $s = 2M\omega$ and with

$$\begin{aligned} Q^2 &= -q^2 \\ \nu &= q\cdot P/M \\ x &= Q^2/(2M\nu) \\ \omega &= 1/(2Mx) \\ K &= M/(kT) \end{aligned} \tag{7.4}$$

where M is the nucleon mass. The calculated $F(\omega)$ together with the experimental data are shown in Fig. 7.1. The gambol-parton energy ε was

$$\varepsilon = ½ m_{g0}v^2 \rightarrow \varepsilon(s) = c^2(m_{g0}\, s + M)/(s + 1) = (sm_{g0} + M)/(s + 1) \tag{7.5}$$

---

[39] See Blaha (2023e) for a calculation using the previous non-scaling Planckian distribution.

where m_{g0} was the gambol-parton mass. This energy parameterization was found to be necessary to obtain a satisfactory range of energies:

$$s = 0 \quad \varepsilon = Mc^2 \quad \text{proton rest mass}$$
$$s = \infty \quad \varepsilon = m_{g0}c^2 \quad \text{gambol rest mass}$$

The constants in the calculation were

$$\begin{aligned} M &= 1.0\ \text{GeV}/c^2 \\ m_{g0} &= M/8 = 0.125\ \text{GeV}/c^2 \\ kT &= 0.328\ M/m_b = 0.0785\ \text{GeV}/c^2 \\ K &= M/\ kT = 12.74 \\ N &= 0.122 \end{aligned} \tag{7.6}$$

The value of m_{g0} corresponds to $x = 0.125$. The effective gambol mass is $m_{g0} = 0.125M = 125\ \text{MeV}/c^2$, which is 1/8 of the proton mass. The corresponding fractionation value is thus

$$s_{effective} = 8 = 1/1.25 \tag{7.7}$$

The appearance of *effective* fractionation 1/8 = (the effective number of gambols in the proton) may be regarded as strongly supportive of fractionation and Limos.

The effective value of kT is 78.5 MeV/c^2. Note $m_{g0} = 3/2$ kT. The normalization is

$$N \cong 3/(2K) \cong m_{g0}/M \tag{7.8}$$

The result is in Fig. 7.1. It shows a relatively close match with the data. Details of this calculation may be found in Blaha (2023e).

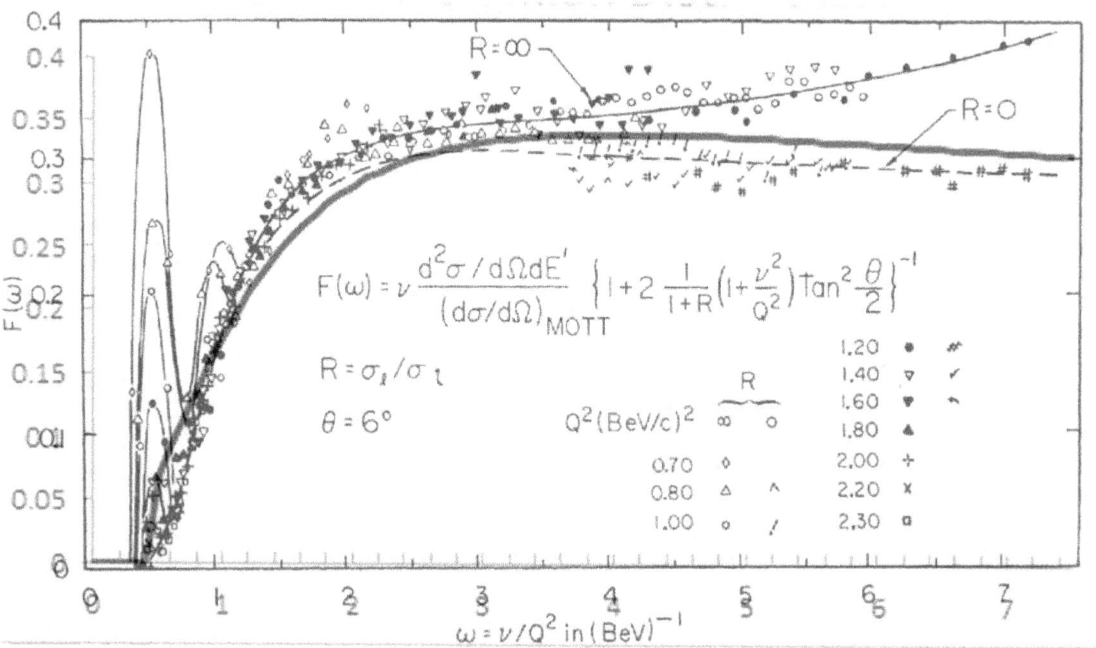

Figure 7.1. Plot of F(ω) = νW2(ω) vs. ω = 2M/x for the non-scaling gambol Planckian distribution and expreimantal data. The solid line is the calculated F(ω) using the non-scaling Planckian distribution of eq. 7.2. The experimental data is from J. D. Bjorken and E. A. Paschos, **185**, 1975 (1969) reprinted with the permission of Physical Review.

7.2 Calculation using the Scaling Planckian Distribution

The *Scaling* Planckian distribution is

$$F(\omega) = U(E) = N(E/kT)^3/(6kT)\exp(-E/kT) \quad (6.11)$$

To determine E and thereby F(ω) we define the quantities followed by comments:

$$M \cong 1.0 \text{ GeV/c}^2 \quad (7.9)$$

Proton mass

$$m_g = M/844242 \cong 1/844242 \text{ GeV/c}^2 \cong 1.4845\cdot10^{-6} \text{ GeV/c}^2$$

Fermion Gambol mass

$$kT = 0.0785 * 3 * M \cong 0.2355 \text{ GeV/c}^2$$

Internal temperature based on Blaha (2023e)

$$N = 0.4$$

Normalization factor for other dynamic effects

$$\omega = s/(2M) \cong s/(2 \text{ GeV/c}^2)$$

$$m = (26382\omega m_g + 1)/(\omega + 1)$$
$$= (844242/32\ \ s/2\ m_g + 1)/(s/2 + 1)$$
$$= (26382 s m_g + 2)/(s + 2)$$

Mass energy of a 32 Gambol Fermion clump with which deep inelastic photons interact. We assume the very small mass of the Gambol Fermions favors interactions in clumps of 32.

$$E = (m\,s + M)/(s + 1) - M$$
$$= (M - m)s/(s + 1)$$

The energy depends on m which has the mass of a clump. A clump's mass is intermediate between the proton mass and the Gambol Fermion mass. It is a sort of "bridge" between these masses that avoids having terms with large differences in mass. The factor (M – m)s sets a scale of interacting clumps.

Fig. 7.2 displays F(ω). The match with the experimental data is impressive. See Fig. 7.3 for the superimposed F(ω) and experimental data (and the non-scaling Planckian results from Blaha (2023e).

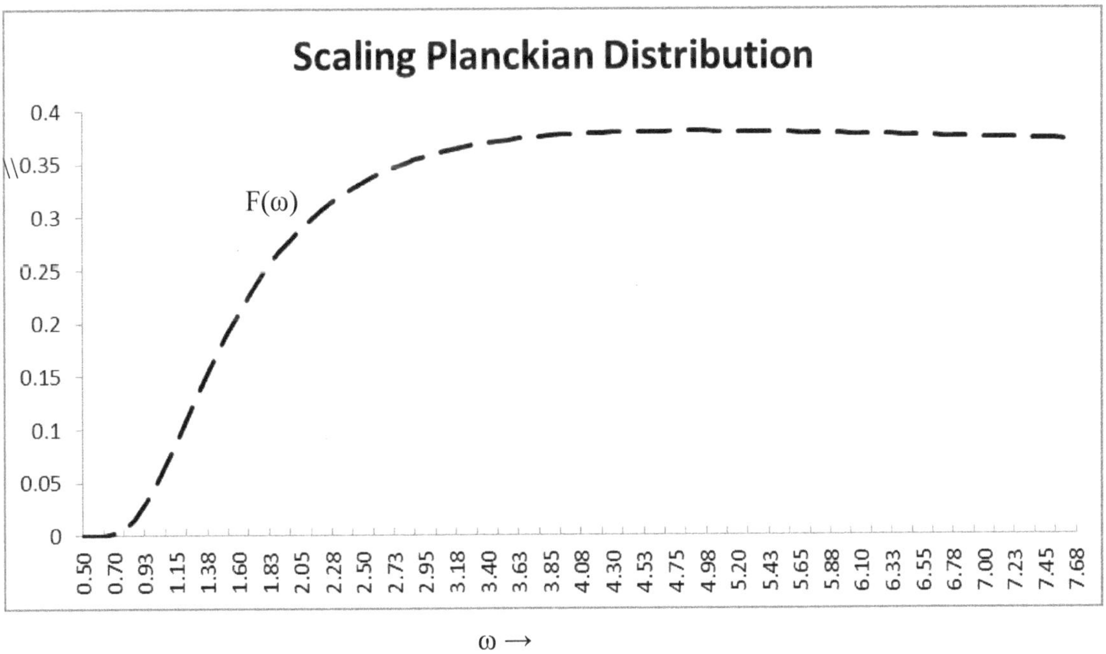

Figure 7.2. Plot of the calculated Scaling Planckian distribution for F(ω) for deep inelastic e-p scattering.

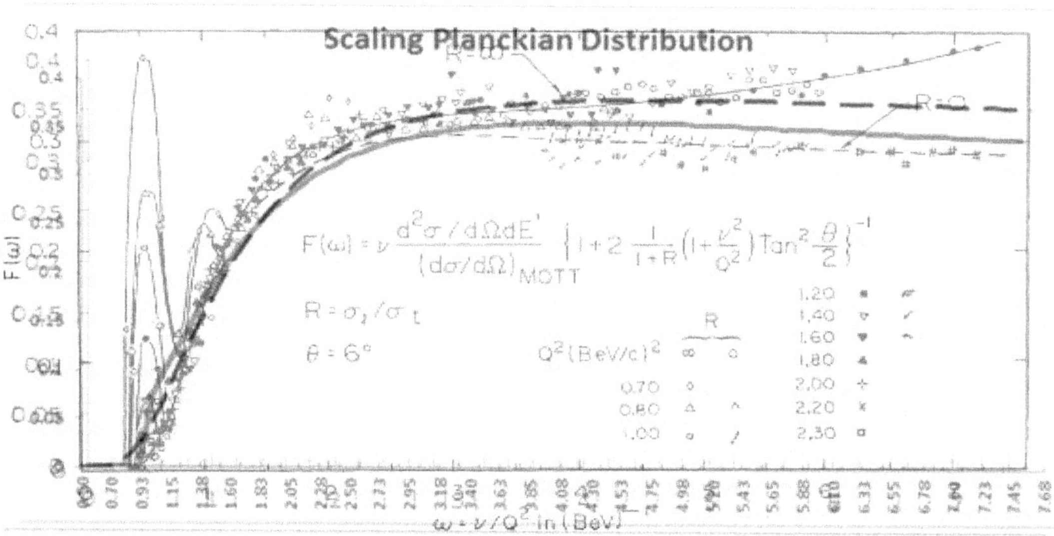

Figure 7.3. Plot of calculated Scaling Planckian distribution for F(ω) for deep inelastic e-p scattering superimposed on Fig. 7.1. The dashed line is the Scaling Planckian distribution of F(ω) from eq. 6.11 with the data of eq. 7.9. An excellent match with the experimental data. It appears better than the previous calculation sisplayed in Fig. 7.1 with a non-scaling Planckian distribution. The large ω part of the dashed line can be adjusted higher or lower by changes in the kT parameter.

8. What Do We Make of Cosmos Theory

Cosmos Theory has a great variety of theoretic features, *and* experimental implications, that are beyond current thinking. It appears to give a complete characterization of Physical Reality. So complete as to raise the question of a close connection to true Reality as we know it.

What shall we make of it? We suggest the reader follow Hamlet's[40] thought, "Absent thee from felicity a while" in examining Cosmos Theory.

[40] Hamlet speaking to Horatio in Shakespeare's *Hamlet.*

REFERENCES

Akhiezer, N. I., Frink, A. H. (tr), 1962, *The Calculus of Variations* (Blaisdell Publishing, New York, 1962).
Bjorken, J. D., Drell, S. D., 1964, *Relativistic Quantum Mechanics* (McGraw-Hill, New York, 1965).
Bjorken, J. D., Drell, S. D., 1965, *Relativistic Quantum Fields* (McGraw-Hill, New York, 1965).
Blaha, S., 1995, *C++ for Professional Programming* (International Thomson Publishing, Boston, 1995).
_______, 1998, *Cosmos and Consciousness* (Pingree-Hill Publishing, Auburn, NH, 1998 and 2002).
_______, 2002, *A Finite Unified Quantum Field Theory of the Elementary Particle Standard Model and Quantum Gravity Based on New Quantum Dimensions™ & a New Paradigm in the Calculus of Variations* (Pingree-Hill Publishing, Auburn, NH, 2002).
_______, 2004, *Quantum Big Bang Cosmology: Complex Space-time General Relativity, Quantum Coordinates™ Dodecahedral Universe, Inflation, and New Spin 0, ½, 1 & 2 Tachyons & Imagyons* (Pingree-Hill Publishing, Auburn, NH, 2004).
______, *2005a, Quantum Theory of the Third Kind: A New Type of Divergence-free Quantum Field Theory Supporting a Unified Standard Model of Elementary Particles and Quantum Gravity based on a New Method in the Calculus of Variations* (Pingree-Hill Publishing, Auburn, NH, 2005).
______, 2005b, *The Metatheory of Physics Theories, and the Theory of Everything as a Quantum Computer Language* (Pingree-Hill Publishing, Auburn, NH, 2005).
______, 2005c, *The Equivalence of Elementary Particle Theories and Computer Languages: Quantum Computers, Turing Machines, Standard Model, Superstring Theory, and a Proof that Gödel's Theorem Implies Nature Must Be Quantum* (Pingree-Hill Publishing, Auburn, NH, 2005).
______, 2006a, *The Foundation of the Forces of Nature* (Pingree-Hill Publishing, Auburn, NH, 2006).
______, 2006b, *A Derivation of ElectroWeak Theory based on an Extension of Special Relativity; Black Hole Tachyons; & Tachyons of Any Spin.* (Pingree-Hill Publishing, Auburn, NH, 2006).
_______, 2007a, *Physics Beyond the Light Barrier: The Source of Parity Violation, Tachyons, and A Derivation of Standard Model Features* (Pingree-Hill Publishing, Auburn, NH, 2007).
______, 2007b, *The Origin of the Standard Model: The Genesis of Four Quark and Lepton Species, Parity Violation, the ElectroWeak Sector, Color SU(3), Three Visible Generations of Fermions, and One Generation of Dark Matter with Dark Energy* (Pingree-Hill Publishing, Auburn, NH, 2007).
______, *2008a, A Direct Derivation of the Form of the Standard Model From GL(16) (Pingree-Hill Publishing, Auburn, NH, 2008).*
______, 2008b, *A Complete Derivation of the Form of the Standard Model With a New Method to Generate Particle Masses Second Edition* (Pingree-Hill Publishing, Auburn, NH, 2008)
______, 2009, *The Algebra of Thought & Reality: The Mathematical Basis for Plato's Theory of Ideas, and Reality Extended to Include A Priori Observers and Space-Time Second Edition* (Pingree-Hill Publishing, Auburn, NH, 2009).
______, 2009a, *Bright Stars, Bright Universe* (Pingree-Hill Publishing, Auburn, NH, 2008)
______, 2010a, *Operator Metaphysics: A New Metaphysics Based on a New Operator Logic and a New Quantum Operator Logic that Lead to a Mathematical Basis for Plato's Theory of Ideas and Reality* (Pingree-Hill Publishing, Auburn, NH, 2010).
______, 2010b, *The Standard Model's Form Derived from Operator Logic, Superluminal Transformations and GL(16)* (Pingree-Hill Publishing, Auburn, NH, 2010).
______, 2010c, *SuperCivilizations: Civilizations as Superorganisms* (McMann-Fisher Publishing, Auburn, NH, 2010).
______, 2011a, *21st Century Natural Philosophy Of Ultimate Physical Reality* (McMann-Fisher Publishing, Auburn, NH, 2011).
______, 2011b, *All the Universe! Faster Than Light Tachyon Quark Starships & Particle Accelerators with the LHC as a Prototype Starship Drive Scientific Edition* (Pingree-Hill Publishing, Auburn, NH, 2011).
______, 2011c, *From Asynchronous Logic to The Standard Model to Superflight to the Stars* (Blaha Research, Auburn, NH, 2011).

______, 2012a, *From Asynchronous Logic to The Standard Model to Superflight to the Stars volume 2: Superluminal CP and CPT, U(4) Complex General Relativity and The Standard Model, Complex Vierbein General Relativity, Kinetic Theory, Thermodynamics* (Blaha Research, Auburn, NH, 2012).
______, 2012b, *Standard Model Symmetries, And Four And Sixteen Dimension Complex Relativity; The Origin Of Higgs Mass Terms* (Blaha Reasearch, Auburn, NH, 2012).
______, 2013a, *Multi-Stage Space Guns, Micro-Pulse Nuclear Rockets, and Faster-Than-Light Quark-Gluon Ion Drive Starships* (Blaha Research, Auburn, NH, 2013).
______, 2013b, *The Bridge to Dark Matter; A New Sibling Universe; Dark Energy; Inflatons; Quantum Big Bang; Superluminal Physics; An Extended Standard Model Based on Geometry* (Blaha Reasearch, Auburn, NH, 2013).
______, 2014a, *Universes and Megaverses: From a New Standard Model to a Physical Megaverse; The Big Bang; Our Sibling Universe's Wormhole; Origin of the Cosmological Constant, Spatial Asymmetry of the Universe, and its Web of Galaxies; A Baryonic Field between Universes and Particles; Megaverse Extended Wheeler-DeWitt Equation* (Blaha Reasearch, Auburn, NH, 2014).
______, 2014b, *All the Megaverse! Starships Exploring the Endless Universes of the Cosmos Using the Baryonic Force* (Blaha Research, Auburn, NH, 2014).
______, 2014c, *All the Megaverse! II Between Megaverse Universes: Quantum Entanglement Explained by the Megaverse Coherent Baryonic Radiation Devices – PHASERs Neutron Star Megaverse Slingshot Dynamics Spiritual and UFO Events, and the Megaverse Microscopic Entry into the Megaverse* (Blaha Research, Auburn, NH, 2014).
______, 2015a, *PHYSICS IS LOGIC PAINTED ON THE VOID: Origin of Bare Masses and The Standard Model in Logic, U(4) Origin of the Generations, Normal and Dark Baryonic Forces, Dark Matter, Dark Energy, The Big Bang, Complex General Relativity, A Megaverse of Universe Particles* (Blaha Research, Auburn, NH, 2015).
______, 2015b, *PHYSICS IS LOGIC Part II: The Theory of Everything, The Megaverse Theory of Everything, U(4)⊗U(4) Grand Unified Theory (GUT), Inertial Mass = Gravitational Mass, Unified Extended Standard Model and a New Complex General Relativity with Higgs Particles, Generation Group Higgs Particles* (Blaha Research, Auburn, NH, 2015).
____, 2015c, *The Origin of Higgs ("God") Particles and the Higgs Mechanism: Physics is Logic III, Beyond Higgs – A Revamped Theory With a Local Arrow of Time, The Theory of Everything Enhanced, Why Inertial Frames are Special, Universes of the Mind* (Blaha Research, Auburn, NH, 2015).
______, 2015d, *The Origin of the Eight Coupling Constants of The Theory of Everything: U(8) Grand Unified Theory of Everything (GUTE), S^8 Coupling Constant Symmetry, Space-Time Dependent Coupling Constants, Big Bang Vacuum Coupling Constants, Physics is Logic IV* (Blaha Research, Auburn, NH, 2015).
______, 2016a, *New Types of Dark Matter, Big Bang Equipartition, and A New U(4) Symmetry in the Theory of Everything: Equipartition Principle for Fermions, Matter is 83.33% Dark, Penetrating the Veil of the Big Bang, Explicit QFT Quark Confinement and Charmonium, Physics is Logic V* (Blaha Research, Auburn, NH, 2016).
______, 2016b, *The Periodic Table of the 192 Quarks and Leptons in The Theory of Everything: The U(4) Layer Group, Physics is Logic VI* (Blaha Research, Auburn, NH, 2016).
______, 2016c, *New Boson Quantum Field Theory, Dark Matter Dynamics, Dark Matter Fermion Layer Mixing, Genesis of Higgs Particles, New Layer Higgs Masses, Higgs Coupling Constants, Non-Abelian Higgs Gauge Fields, Physics is Logic VII* (Blaha Research, Auburn, NH, 2016).
______, 2016d, *Unification of the Strong Interactions and Gravitation: Quark Confinement Linked to Modified Short-Distance Gravity; Physics is Logic VIII* (Blaha Research, Auburn, NH, 2016).
______, 2016e, *MoND: Unification of the Strong Interactions and Gravitation II, Quark Confinement Linked to Large-Scale Gravity, Physics is Logic IX* (Blaha Research, Auburn, NH, 2016).
______, 2016f, *CQ Mechanics: A Unification of Quantum & Classical Mechanics, Quantum/Semi-Classical Entanglement, Quantum/Classical Path Integrals, Quantum/Classical Chaos* (Blaha Research, Auburn, NH, 2016).
______, 2016g, *GEMS Unified Gravity, ElectroMagnetic and Strong Interactions: Manifest Quark Confinement, A Solution for the Proton Spin Puzzle, Modified Gravity on the Galactic Scale* (Pingree Hill Publishing, Auburn, NH, 2016).
______, 2016h, *Unification of the Seven Boson Interactions based on the Riemann-Christoffel Curvature Tensor* (Pingree Hill Publishing, Auburn, NH, 2016).
______, 2017a, *Unification of the Eleven Boson Interactions based on 'Rotations of Interactions'* (Pingree Hill Publishing, Auburn, NH, 2017).
______, 2017b, *The Origin of Fermions and Bosons, and Their Unification* (Pingree Hill Publishing, Auburn, NH, 2017).
______, 2017c, *Megaverse: The Universe of Universes* (Pingree Hill Publishing, Auburn, NH, 2017).
______, 2017d, *SuperSymmetry and the Unified SuperStandard Model* (Pingree Hill Publishing, Auburn, NH, 2017).

______, 2017e, *From Qubits to the Unified SuperStandard Model with Embedded SuperStrings: A Derivation* (Pingree Hill Publishing, Auburn, NH, 2017).
______, 2017f, *The Unified SuperStandard Model in Our Universe and the Megaverse: Quarks, ...* , (Pingree Hill Publishing, Auburn, NH, 2017).
______, 2018a, *The Unified SuperStandard Model and the Megaverse SECOND EDITION A Deeper Theory based on a New Particle Functional Space that Explicates Quantum Entanglement Spookiness (Volume 1)* (Pingree Hill Publishing, Auburn, NH, 2018).
______, 2018b, *Cosmos Creation: The Unified SuperStandard Model, Volume 2, SECOND EDITION* (Pingree Hill Publishing, Auburn, NH, 2018).
______, 2018c, *God Theory* (Pingree Hill Publishing, Auburn, NH, 2018).
______, 2018d, *Immortal Eye: God Theory: Second Edition* (Pingree Hill Publishing, Auburn, NH, 2018).
______, 2018e, *Unification of God Theory and Unified SuperStandard Model THIRD EDITION* (Pingree Hill Publishing, Auburn, NH, 2018).
______, 2019a, *Calculation of: QED α = 1/137, and Other Coupling Constants of the Unified SuperStandard Theory* (Pingree Hill Publishing, Auburn, NH, 2019).
______, 2019b, *Coupling Constants of the Unified SuperStandard Theory SECOND EDITION* (Pingree Hill Publishing, Auburn, NH, 2019).
______, 2019c, *New Hybrid Quantum Big_Bang–Megaverse_Driven Universe with a Finite Big Bang and an Increasing Hubble Constant* (Pingree Hill Publishing, Auburn, NH, 2019).
______, 2019d, *The Universe, The Electron and The Vacuum* (Pingree Hill Publishing, Auburn, NH, 2019).
______, 2019e, *Quantum Big Bang – Quantum Vacuum Universes (Particles)* (Pingree Hill Publishing, Auburn, NH, 2019).
______, 2019f, *The Exact QED Calculation of the Fine Structure Constant Implies ALL 4D Universes have the Same Physics/Life Prospects* (Pingree Hill Publishing, Auburn, NH, 2019).
______, 2019g, *Unified SuperStandard Theory and the SuperUniverse Model: The Foundation of Science* (Pingree Hill Publishing, Auburn, NH, 2019).
______, 2020a, *Quaternion Unified SuperStandard Theory (The QUeST) and Megaverse Octonion SuperStandard Theory (MOST)* (Pingree Hill Publishing, Auburn, NH, 2020).
______, 2020b, *United Universes Quaternion Universe - Octonion Megaverse* (Pingree Hill Publishing, Auburn, NH, 2020).
______, 2020c, *Unified SuperStandard Theories for Quaternion Universes & The Octonion Megaverse* (Pingree Hill Publishing, Auburn, NH, 2020).
______, 2020d, *The Essence of Eternity: Quaternion & Octonion SuperStandard Theories* (Pingree Hill Publishing, Auburn, NH, 2020).
______, 2020e, *The Essence of Eternity II* (Pingree Hill Publishing, Auburn, NH, 2020).
______, 2020f, *A Very Conscious Universe* (Pingree Hill Publishing, Auburn, NH, 2020).
______, 2020g, *Hypercomplex Universe* (Pingree Hill Publishing, Auburn, NH, 2020).
______, 2020h, *Beneath the Quaternion Universe* (Pingree Hill Publishing, Auburn, NH, 2020).
______, 2020i, *Why is the Universe Real? From Quaternion & Octonion to Real Coordinates* (Pingree Hill Publishing, Auburn, NH, 2020).
______, 2020j, *The Origin of Universes: of Quaternion Unified SuperStandard Theory (QUeST); and of the Octonion Megaverse (UTMOST)* (Pingree Hill Publishing, Auburn, NH, 2020).
______, 2020k, *The Seven Spaces of Creation: Octonion Cosmology* (Pingree Hill Publishing, Auburn, NH, 2020).
______, 2020l, *From Octonion Cosmology to the Unified SuperStandard Theory of Particles* (Pingree Hill Publishing, Auburn, NH, 2020).
______, 2021a, *Pioneering the Cosmos* (Pingree Hill Publishing, Auburn, NH, 2021).
______, 2021b, *Pioneering the Cosmos II* (Pingree Hill Publishing, Auburn, NH, 2021).
______, 2021c, *Beyond Octonion Cosmology* (Pingree Hill Publishing, Auburn, NH, 2021).
______, 2021d, *Universes are Particles* (Pingree Hill Publishing, Auburn, NH, 2021).
______, 2021e, *Octonion-like dna-based life, Universe expansion is decay, Emerging New Physics* (Pingree Hill Publishing, Auburn, NH, 2021).
______, 2021f, *The Science of Creation New Quantum Field Theory of Spaces* (Pingree Hill Publishing, Auburn, NH, 2021).
______, 2021g, *Quantum Space Theory With Application to Octonion Cosmology & Possibly To Fermionic Condensed Matter* (Pingree Hill Publishing, Auburn, NH, 2021).
______, 2021h, *21st Century Natural Philosophy of Octonion Cosmology , and Predestination, Fate, and Free Will* (Pingree Hill Publishing, Auburn, NH, 2021).

______, 2021i, *Beyond Octonion Cosmology II : Origin of the Quantum; A New Generalized Field Theory (GiFT); A Proof of the Spectrum of Universes; Atoms in Higher Universes* (Pingree Hill Publishing, Auburn, NH, 2021).
______, 2021j, *Integration of General Relativity and Quantum Theory: Octonion Cosmology, GiFT, Creation/Annihilation Spaces CASe, Reduction of Spaces to a Few Fermions and Symmetries in Fundamental Frames* (Pingree Hill Publishing, Auburn, NH, 2021).
______, 2022a, *New View of Octonion Cosmology Based on the Unification of General Relativity and Quantum Theory* (Pingree Hill Publishing, Auburn, NH, 2022).
______, 2022b, *The Dust Beneath Hypercomplex Cosmology* (Pingree Hill Publishing, Auburn, NH, 2022).
______, 2022c, *Passing Through Nature to Eternity: ProtoCosmos, HyperCosmos, Unified SuperStandard Theory* (Pingree Hill Publishing, Auburn, NH, 2022).
______, 2022d, *HyperCosmos Fractionation and Fundamental Reference Frame Based Unification: Particle Inner Space Basis of Parton and Dual Resonance Models* (Pingree Hill Publishing, Auburn, NH, 2022).
______, 2022e, *A New UniDimension ProtoCosmos and SuperString F-Theory Relation to the HyperCosmos* (Pingree Hill Publishing, Auburn, NH, 2022).
______, 2022f, *The Cosmic Panorama: ProtoCosmos, HyperCosmos, Unified SuperStandard Theory (UST) Derivation* (Pingree Hill Publishing, Auburn, NH, 2022).
_____, 2022g, *Ultimate Origin: ProtoCosmos and HyperCosmos* (Pingree Hill Publishing, Auburn, NH, 2022).
______, 2023a, *UltraUnification and the Generation of the Cosmos* (Pingree Hill Publishing, Auburn, NH, 2023).
______, 2023b, *God and and Cosmos Theory* (Pingree Hill Publishing, Auburn, NH, 2023).
______, 2023c, *A New Completely Geometric SU(8) Cosmos Theory; New PseudoFermion Fields; Fibonacci-like Dimension Arrays; Ramsey Number Approximation* (Pingree Hill Publishing, Auburn, NH, 2023).
______, 2023d, *Newton's Apple is Now the Fermion* (Pingree Hill Publishing, Auburn, NH, 2023).
______, 2023e,*Cosmos Theory: The Sub-Particle Gambol Model* (Pingree Hill Publishing, Auburn, NH, 2023).
_____, 2024a, *Cosmos-Universe-Particle-Gambol Theory* (Pingree Hill Publishing, Auburn, NH, 2024).
_____, 2024b, *Fractal Cosmos Theory* (Pingree Hill Publishing, Auburn, NH, 2024).
_____, 2024c, *Fractal Cosmic Curve: Tensor-Based CosmosTheory* (Pingree Hill Publishing, Auburn, NH, 2024).
_____, 2024d, *The Eternal Form of Cosmos Theory* (Pingree Hill Publishing, Auburn, NH, 2024).
_____, 2024e, *The Eternal Form of Cosmos Theory Third Edition* (Pingree Hill Publishing, Auburn, NH, 2024).
_____, 2024f, *Fundamental Constants of Cosmos Theory and The Standard Model* (Pingree Hill Publishing, Auburn, NH, 2024).
_____, 2024g, *Quark, Lepton, W and Z Masses of Cosmos Theory and The Standard Model* (Pingree Hill Publishing, Auburn, NH, 2024).
_____, 2024h, *Geometric Cosmos Geometric Universe* (Pingree Hill Publishing, Auburn, NH, 2024).
_____, 2024i, *Particles and Universes of Cosmos Theory* (Pingree Hill Publishing, Auburn, NH, 2024).
_____, 2024j, *Unification of the Subluminal and the Superluminal in Cosmos Theory* (Pingree Hill Publishing, Auburn, NH, 2024).
_____, 2024k, *The Dawn of Dynamic Cosmos Dimension Arrays* (Pingree Hill Publishing, Auburn, NH, 2024).
_____, 2024ℓ, *Structure and Dynamics of Cosmos Theory and the Unified SuperStandard Theory* (Pingree Hill Publishing, Auburn, NH, 2024).
_____, 2025a, *Black Holes, White Holes, and Superluminal Starship* (Pingree Hill Publishing, Auburn, NH, 2025).
_____, 2025b, *The Sequence Theory Foundation of Cosmos Theory and Gambol Theory* (Pingree Hill Publishing, Auburn, NH, 2025).
____, 2025c, *BEYOND TIME AND SPACE Fermion Strata & Consistency Conditions for Sequences, Cosmos Spaces and Gambols* (Pingree Hill Publishing, Auburn, NH, 2025).
____, 2025d, *The C o s m o s as a Physics Organism* (Blaha Research, Auburn, NH, 2025).
____, 2025e, *The Root of Reality* (Blaha Research, Auburn, NH, 2025).
____, 2025f, *The Fractal Cosmos* (Blaha Research, Auburn, NH, 2025).
____, 2025g, *The Fractal Cosmos II* (Blaha Research, Auburn, NH, 2025).
Eddington, A. S., 1952, *The Mathematical Theory of Relativity* (Cambridge University Press, Cambridge, U.K., 1952).
Fant, Karl M., 2005, *Logically Determined Design: Clockless System Design With NULL Convention Logic* (John Wiley and Sons, Hoboken, NJ, 2005).
Feinberg, G. and Shapiro, R., 1980, *Life Beyond Earth: The Intelligent Earthlings Guide to Life in the Universe* (William Morrow and Company, New York, 1980).
Gelfand, I. M., Fomin, S. V., Silverman, R. A. (tr), 2000, *Calculus of Variations* (Dover Publications, Mineola, NY, 2000).

Giaquinta, M., Modica, G., Souchek, J., 1998, *Cartesian Coordinates in the Calculus of Variations* Volumes I and II (Springer-Verlag, New York, 1998).
Giaquinta, M., Hildebrandt, S., 1996, *Calculus of Variations* Volumes I and II (Springer-Verlag, New York, 1996).
Gradshteyn, I. S. and Ryzhik, I. M., 1965, *Table of Integrals, Series, and Products* (Academic Press, New York, 1965).
Heitler, W., 1954, *The Quantum Theory of Radiation* (Claendon Press, Oxford, UK, 1954).
Huang, Kerson, 1992, *Quarks, Leptons & Gauge Fields 2nd Edition* (World Scientific Publishing Company, Singapore, 1992).
Jost, J., Li-Jost, X., 1998, *Calculus of Variations* (Cambridge University Press, New York, 1998).
Kaku, Michio, 1993, *Quantum Field Theory*, (Oxford University Press, New York, 1993).
Kirk, G. S. and Raven, J. E., 1962, *The Presocratic Philosophers* (Cambridge University Press, New York, 1962).
Landau, L. D. and Lifshitz, E. M., 1987, *Fluid Mechanics 2nd Edition*, (Pergamon Press, Elmsford, NY, 1987).
Rescher, N., 1967, *The Philosophy of Leibniz* (Prentice-Hall, Englewood Cliffs, NJ, 1967).
Riesz, Frigyes and Sz.-Nagy, Béla, 1990, *Functional Analysis* (Dover Publications, New York, 1990).
Sakurai, J. J., 1964, *Invariance Principles and Elementary Particles* (Princeton University Press, Princeton, NJ, 1964).
Weinberg, S., 1972, *Gravitation and Cosmology* (John Wiley and Sons, New York, 1972).
Weinberg, S., 1995, *The Quantum Theory of Fields Volume I* (Cambridge University Press, New York, 1995).

INDEX

About the Author

Stephen Blaha is a well-known Physicist and Man of Letters with interests in Science, Society and civilization, the Arts, and Technology. He had an Alfred P. Sloan Foundation scholarship in college. He received his Ph.D. in Physics from Rockefeller University. He has served on the faculties of several major universities. He was also a Member of the Technical Staff at Bell Laboratories, a manager at the Boston Globe Newspaper, a Director at Wang Laboratories, and President of Blaha Software Inc. and of Janus Associates Inc. (NH).

Among other achievements he was a co-discoverer of the "r potential" for heavy quark binding developing the first (and still the only demonstrable) non-Aeolian gauge theory with an "r" potential; first suggested the existence of topological structures in superfluid He-3; first proposed Yang-Mills theories would appear in condensed matter phenomena with non-scalar order parameters; first developed a grammar-based formalism for quantum computers and applied it to elementary particle theories; first developed a new form of quantum field theory without divergences (thus solving a major 60 year old problem that enabled a unified theory of the Standard Model and Quantum Gravity without divergences to be developed); first developed a formulation of complex General Relativity based on analytic continuation from real space-time; first developed a generalized non-homogeneous Robertson-Walker metric that enabled a quantum theory of the Big Bang to be developed without singularities at t = 0; first generalized Cauchy's theorem and Gauss' theorem to complex, curved multi-dimensional spaces; received Honorable Mention in the Gravity Research Foundation Essay Competition in 1978; first developed a physically acceptable theory of faster-than-light particles; first derived a composition of extremums method in the Calculus of Variations; first quantitatively suggested that inflationary periods in the history of the universe were not needed; first proved Gödel's Theorem implies Nature must be quantum; provided a new alternative to the Higgs Mechanism, and Higgs particles, to generate masses; first showed how to resolve logical paradoxes including Gödel's Undecidability Theorem by developing Operator Logic and Quantum Operator Logic; first developed a quantitative harmonic oscillator-like model of the life cycle, and interactions, of civilizations; first showed how equations describing superorganisms also apply to civilizations. A recent book shows his theory applies successfully to the past 14 years of history and to *new* archaeological data on Andean and Mayan civilizations as well as Early Anatolian and Egyptian civilizations.

He first developed an axiomatic derivation of the form of The Standard Model from geometry – space-time properties – The Unified SuperStandard Model. It unifies all the known forces of Nature. It also has a Dark Matter sector that includes a Dark ElectroWeak sector with Dark doublets and Dark gauge interactions. It uses quantum coordinates to remove infinities that crop up in most interacting quantum field theories and additionally to remove the infinities that appear in the Big Bang and generate inflationary growth of the universe. It shows gravity has a MOND-like form without sacrificing Newton's Laws. It relates the interactions of the MOND-like sector of gravity with the r-potential of Quark Confinement. The axioms of the theory lead to the question of their origin. We suggest in the preceding edition of this book it can be attributed to an entity with God-like properties. We explore these properties in "God Theory" and show they predict that the Cosmos exists forever although individual universes (or incarnations of our universe) "come and go." Several other important results emerge from God Theory such a functionally triune God. The Unified SuperStandard Theory has many other important parts described in the Current Edition of *The Unified SuperStandard Theory* and expanded in subsequent volumes.

Blaha has had a major impact on a succession of elementary particle theories: his Ph.D. thesis (1970), and papers, showed that quantum field theory calculations to all orders in ladder approximations could not give scaling deep inelastic electron-nucleon scattering. He later showed the eigenvalue equation for the fine structure constant α in Johnson-Baker-Willey QED had a zero at $\alpha = 1$ not 1/137 by solving the Schwinger-Dyson equations to all orders in an approximation that agreed with exact results to 4th order in α thus ending interest in this theory. In 1979 at Professor Ken Johnson's (MIT) suggestion he calculated the proton-neutron mass difference in the MIT bag model and found the result had the wrong sign reducing interest in the bag model. These results all appear in Physical Review papers. In the 2000's he repeatedly pointed out the shortcomings of SuperString theory and showed that The Standard Model's form could be derived from space-time geometry by an extension of Lorentz transformations to faster than light transformations. This deeper space-time basis greatly increases the possibility that it is part of THE fundamental theory. Recently, Blaha showed that the Weak interactions differed significantly from the Strong, electromagnetic and gravitation interactions in important respects while these interactions had similar features, and suggested that ElectroWeak theory, which is essentially a glued union of the Weak interactions and Electromagnetism, possibly modulo unknown Higgs particle features, be replaced by a unified theory of the other interactions combined with a stand-alone Weak interaction theory. Blaha also showed that, if Charmonium calculations are taken seriously, the Strong interaction coupling constant is only a factor of five larger than the electromagnetic coupling constant, and thus Strong interaction perturbation theory would make sense and yield physically meaningful results.

In graduate school (1965-71) he wrote substantial papers in elementary particles and group theory: The Inelastic E- P Structure Functions in a Gluon Model. Phys. Lett. B40:501-502,1972; Deep-Inelastic E-P Structure Functions In A Ladder Model With Spin 1/2 Nucleons, Phys.Rev. D3:510-523,1971; Continuum Contributions To The Pion Radius, Phys. Rev. 178:2167-2169,1969; Character Analysis of U(N) and SU(N), J. Math. Phys. 10, 2156 (1969); and The Calculation of the Irreducible Characters of the Symmetric Group in Terms of the Compound Characters, (Published as Blaha's Lemma in D. E. Knuth's book: *The Art of Computer Programming Vols. 1 – 4*).

In the early 1980's Blaha was also a pioneer in the development of UNIX for financial, scientific and Internet applications: benchmarked UNIX versions showing that block size was critical for UNIX performance, developing financial modeling software, starting database benchmarking comparison studies, developing Internet-like UNIX networking (1982) and developing a hybrid shell programming technique (1982) that was a precursor to the PERL programming language. He was also the manager of the AT&T ten-year future products development database. His work helped lead to commercial UNIX on computers such as Sun Micros, IBM AIX minis, and Apple computers.

In the 1980's he pioneered the development of PC Desktop Publishing on laser printers and was nominated for three "Awards for Technical Excellence" in 1987 by PC Magazine for PC software products that he designed and developed.

Recently he has developed a theory of Megaverses – actual universes of which our universe is one – with quantum particle-like properties based on the Wheeler-DeWitt equation of Quantum Gravity. He has developed a theory of a baryonic force, which had been conjectured many years ago, and estimated the strength of the force based on discrepancies in measurements of the gravitational constant G. This force, operative in D-dimensional space, can be used to escape from our universe in "uniships" which are the equivalent of the faster-than-light starships proposed in the author's earlier books. Thus travel to other universes, as well as to other stars is possible.

Blaha also considered the complexified Wheeler-DeWitt equation and showed that its limitation to real-valued coordinates and metrics generated a Cosmological Constant in the Einstein equations.

The author has also recently written a series of books on the serious problems of the United States and their solution as well as a book on the decline of Mankind that will follow from current social and genetic trends in Mankind.

In the past twenty years Dr. Blaha has written over 80 books on a wide range of topics. Some recent major works are: *From Asynchronous Logic to The Standard Model to Superflight to the Stars*, *All the Universe!*, *SuperCivilizations: Civilizations as Superorganisms*, *America's Future: an Islamic Surge, ISIS, al Qaeda, World Epidemics, Ukraine, Russia-China Pact, US Leadership Crisis*, *The Rises and Falls of Man – Destiny – 3000 AD: New Support for a Superorganism MACRO-THEORY of CIVILIZATIONS From CURRENT WORLD TRENDS and NEW Peruvian, Pre-Mayan, Mayan, Anatolian, and Early Egyptian Data, with a Projection to 3000 AD,* and *Mankind in Decline: Genetic Disasters, Human-Animal Hybrids, Overpopulation, Pollution, Global Warming, Food and Water Shortages, Desertification, Poverty, Rising Violence, Genocide, Epidemics, Wars, Leadership Failure.*

He has taught approximately 4,000 students in undergraduate, graduate, and postgraduate corporate education courses primarily in major universities, and large companies and government agencies.

He developed a quantum theory, The Unified SuperStandard Theory (UST), which describes elementary particles in detail without the difficulties of conventional quantum field theory. He found that the internal symmetries of this theory could be exactly derived from an octonion theory called QUeST. He further found that another octonion theory (UTMOST) describes the Megaverse. It can hold QUeST universes such as our own universe. It has an internal symmetry structure which is a superset of the QUeST internal symmetries.

Recently he developed Octonion Cosmology. He replaced it with HyperCosmos theory, which has significantly better features. He developed a fractionalization process for dimensions, particles and symmetry groups. He also described transformation that reduced particles and dimensions to a far more compact form. He also developed a precursor theory ProtoCosmos that leads to the HyperCosmos.

The author showed that space-time and Internal Symmetries can be unified in any of the ten HyperCosmos spaces in their associated HyperUnification spaces. The combined set of HyperUnification spaces enable all HyperCosmos dimensions to be obtained by a General Relativistic transformation from one primordial dimension in the 42 space-time dimension unified HyperUnification space.

At present the author devel;oped the Cosmos Theory that incorporates ProtoCosmos Theory, HyperCosmos Theory, Limos Theory, Second Kind HyperCosmos Theory and HyperUnification Spaces. He has introduced PseudoFermion wave functions and theory, He has related Cosmos Theory to Regge trajectories of spaces, parton theory, Veneziano amplitudes, Fibonacci numbers and Ramsey numbers. He has calculated an approximation to the difficult R(n,n) Ramsey numbers.

He has developed a Gambol Model that successfully accounts for e-p deep inelastic scattering, fundamental particle resonances, hadron scattering, and the inner structure of particles based on confinement through Casimir forces of ideal gambol gases. The Gambol Planckian Distribution was derived.

He has applied the Gambol Model to particles, universes, and the Cosmos of universes. He showed that the Cosmos may have a distribution of 23 universes corresponding to various Cosmos spaces.

Recently he showed that Cosmos Theory follows from the number of independent asymmetric tensors in a dimension r. He also showed the close parallel between the form of γ-matrices and Cosmos Theory dimension arrays. The closeness suggested that dimension arrays have the same importance as γ-matrices for fermions.

He demonstrated that the pressure of fermions within a space of dimension r balances the Casimir vacuum energy force for 18 dimensions. He showed that $2e\pi$ = 17.02 marks the critical point where pressure balances Casimir force, which implies r = 18 is the highest dimension Physical Cosmos space. The dimension $2e\pi$ appears to set the approximate dimension for Cosmos spaces with dimension array size $2^{r+4} \cong (17.02/8)^{r+4}. \cong (e\pi/4)^{r+4}. \cong 2.13^{r+4}$.

Now he has found the sequences of Coupling Constant values and fermion masses in the Standard Model and UST. This book unifies the Cosmos spaces spectrum, the coupling constant spectrum and the fermion mass sequences in a fundamental scaling differential equation.

This book shows the fractal basis of Cosmos Theory.

www.ingramcontent.com/pod-product-compliance
Lightning Source LLC
LaVergne TN
LVHW071339210926
848486LV00003B/185